Oil and Gas Production
in Nontechnical Language

Oil and Gas Production
in Nontechnical Language

Martin Raymond and William Leffler

Copyright ©2006 by
PennWell Corporation
1421 South Sheridan Road
Tulsa, Oklahoma 74112-6600 USA

800.752.9764
+1.918.831.9421
sales@pennwell.com
www.pennwellbooks.com
www.pennwell.com

Managing Editor: Marla Patterson
Production /Operations Manager: Traci Huntsman
Production Manager: Robin Remaley
Assistant Editor: Amethyst Hensley
Production Assistant: Amanda Seiders
Book Designer: Clark Bell

Library of Congress Cataloging-in-Publication Data Available on Request

Raymond, Martin and William Leffler
Petroleum Production in Nontechnical Language, Third Edition
ISBN 1-59370-052-0

Printed in the United States of America

1 2 3 4 5   10 09 08 07 06

Contents

Foreword

I often have thought that what oil and gas companies actually do to discover and produce hydrocarbons must be a mystery to many people outside the E&P community. Nor do they fully appreciate the large sums of capital put at risk. They may not even realize that it is in production operations that all the upstream efforts of those companies turn into revenue, into money.

Still, that simple view fails to account for the value generated by almost countless numbers of people that provide the services, materials, and capital vital to successful E&P ventures. As authors Bill Leffler and Martin Raymond point out in this book, more than 80% of the money that oil and gas companies spend goes not to their own engineers, scientists, and operating staff, but to service companies and suppliers. And that doesn't even count the support services within the oil and gas companies.

I don't doubt that in those specialized companies and support groups there are many, many people who want to and ought to know more about the processes of producing oil and gas. And I am also convinced that if they did, both they and the companies they support would achieve continuously increasing levels of efficiency and effectiveness.

This book by Leffler and Raymond is a broad leap across the gap between the mysteries of production operations and the need for better understanding by those who help make it happen. This book won't tell engineers and operating people how to do their jobs, but it will make clear to people who have to deal with them what those engineers and operating people are trying to achieve—and why.

Martin Raymond's long career in production and Bill Leffler's broad oil and gas background and credentials as a writer make them the right team to create this essential book.

John F. Bookout
President and CEO (retired),
Shell Oil Company

Preface

Who?

To petroleum engineers and geologists, the basics of oil and gas production are virtually second nature. That's what they do.

But what about the rest of the world—the mud salesman, the information technology specialist, the environmentalist, the accountant, the facilities engineer, the seismic crew member, the...well, you get it. All these people have to deal with petroleum engineers and geologists, providing the goods and services. How do they get a grip on the challenges of extracting oil and gas from the ground? How do they relate announcements about new technologies and innovations to what their clients are currently doing? And how does another group, those abruptly thrust into the industry—a landowner, a royalty-interest owner, or an incredibly lucky heir—catch up?

We wrote this book with all those people in mind. Some are engineering graduates. Many have only a vaguely related technical education. Others don't even have that arrow in their quivers. So this book attempts to reduce the technology to understandable prose. Oh, there are one or two sections that have formulas, but that's all. There may be complicated charts and diagrams, but every one has an easy explanation—even though we acknowledge that production is a complicated business.

What?

The meat of this book is in the second two-thirds. But at any proper meal, an appetizer, soup, and salad should come first. That's why the first third has the "upstream part" of the upstream—some geology and geophysics, some legal stuff, and drilling. All the petroleum engineers and the geologists had to learn it before they could function. Without it, the business of production would remain a mystery.

Most of the last two-thirds of this book deal with the theory and operations that take place at the lease

- describing what's in the subsurface

- how it reacts when tapped by a well

- how to make the commodity saleable

Toward the end, two short chapters deal with the people running the show and how they decide what part of the business would make them the most money.

Where?

The scale and scope of oil and gas production cover both the world's largest field, Ghawar, in Saudi Arabia, which produces more than four million barrels a day, as well as a one-stripper-well field in West Texas averaging two barrels a day. Most of the world's oil fields are more like the one in West Texas than the one in Saudi Arabia. Look at the

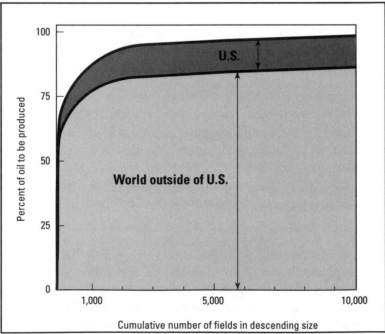

Fig. P–1. Ultimate oil recovered in the world by field size. Only a few hundred fields account for most of the oil that will ever be produced. A seemingly endless number of small fields make up the remaining share.

distribution of oil reserves by field size in the figure on this page. Nearly 90% of the oil to be produced from fields already discovered will come from only 10% of all fields. But still, the other 90% of the fields represent ongoing and nearly endless business activities and opportunities for potential readers of this book.

How?

While we labored over the prose and illustrations of this book, our wives patiently (almost always) let each of us huddle for long, uncommunicative hours. Without such tolerance, a less satisfying product would have emerged. Our thanks go to them. Ask any well-married author and you'll get the same story.

Besides that great boon, we had valuable help from a number of experts, Laura Raymond, Sam Peppiatt, Frank Wolfe, Bob Henly, and Bob Glenn. Pat Raymond read and critiqued the manuscript and gave a us a perspective from our intended audience. Always we interpreted what everyone said, so we have to take sole responsibility for what you see and read here.

M.S.R. and W.L.L.

List of Figures

1

How Did We Get Here?
The History of Production

What is history but a fable agreed upon?

—Napoleon Bonaparte (1769–1821)

Oily Beginnings

In 1859, a character with the unlikely name of Uncle Billy reached down with a tin scoop into the hole he had just drilled, drew a sample of the fluids, and started the era of petroleum production. He smelled it; he tasted it; he rubbed it between his fingers and then giddily dispatched his gofer to town to notify Colonel Edwin Drake he had just struck oil.

Most petro-historians start with the travails of the colonel that led up to spudding the Drake well. But this is a book about production, not drilling, so it should start at the moment when he arrived on the scene, flushed with excitement on the news that his driller, Billy Smith, had hit an oil deposit 69½ feet below ground, near the oil seeps at Titusville, Pennsylvania. Drake dragged an iron water pump from the equipment pile, lowered it into the well with sections of threaded shaft. He rigged the pump handle to the oscillating arm that Uncle Billy used to drive his cable tool drilling rig and began the world's first "modern" oil production—into a metal washtub.

The opportunities to improve Drake's simplistic operation abounded. Over the next century and a half, engineers have tirelessly devised ingenious ways to move the oil (and eventually gas) from its resting place in the ground to the pipelines and trucks that hauled it to market—including the pumps, the shafts, the power units, the wellbore itself, and all the surface equipment to make the oil and gas marketable. Still, the appearance of many of today's well sites isn't remarkably different from those that sprung up in western Pennsylvania over the ensuing 10 years.

On his bare-bones budget, Drake employed the cheapest surface equipment he could find. In a quick upgrade, he switched his production from a washtub to a wooden fish oil container. From there, he filled 50-gallon wooden barrels and had them hauled away by horse-drawn wagons to market.

Eventually, Drake and his contemporaries switched to larger wooden-staved storage tanks. For decades, tank maintenance often meant pounding the tops of the staves to tighten the walls and reduce leakage. At the more prolific wells, oil was simply pumped into a sump (fig. 1–1) until it could be hauled away, a practice that lasted well into the 20th century. By the 1880s, riveted and bolted wrought-iron tanks began to appear at the well sites, but many of those were susceptible to leaks, lightning strikes, and fires (fig. 1–2). Welded steel tanks didn't become popular until 60 years after Drake filled his first fish oil container.

Fig. 1–1. Oil sump at Spindletop, ca. 1902. Oil sumps were open to the air, and much of the gasoline and naphtha just evaporated. Producers didn't mind—the internal combustion engine had not been invented yet. They were after the kerosene. Courtesy of Spindletop Museum.

Fig. 1–2. Burning tank at Spindletop, ca. 1902. Courtesy of Spindletop Museum.

The stampede of oilmen that followed Drake soon found that producing oil was not as easy as the centuries-old technology of producing water. As they drilled deeper, they found the old pitcher pump, the kind that had to be primed to pump water, could no longer pull enough suction. They replaced it with a plunger-type pump that had two ball valves: one at the top, which opened on the downstroke and closed on the upstroke, and the other at the bottom, which did the reverse. All worked well with the plunger pump unless they encountered oil with a lot of dissolved gas, which sometimes caused vapor lock in the pump chamber.

The tinkerers rose to the occasion and stormed the U.S. Patent Office with clever designs. In 1866, for example, R. Cornelius put a wide spot in the tubing just above the top of the plunger stroke to allow the gas to slip out and the column of oil to displace the accumulated gas in the chamber. S. H. Early devised a slide valve to let the gas escape up the annulus between the tubing and the casing. Countless variations and continuous improvements on the basic plunger pump have been made since then.

With the advent of electricity at the well site, bottom-hole, submersible electric pumps became practical. H. W. Pickett started selling one in 1894 that featured a "Yankee screwdriver," a set of right and left threads on the same shaft. This clever mechanism allowed the shaft to rotate in opposite directions during the up- and downstrokes. A myriad of rotary and reciprocating electric motors followed, driving rotary pumps, reciprocating pumps, and hydraulic pumps in endlessly conceived designs and combinations.

A Case for Casing

Installing casing (cylindrical, iron or steel pipe), by the way, quickly became a standard technique. Colonel Drake had the insight to put surface casing down about 32 feet, pounding soft, iron three-inch-diameter pipe with a white oak battering ram. The remaining 45 feet remained open hole. Nagging problems with cave-ins, water intrusion, and gas soon motivated drillers to run casing down the wellbore to just above the producing sands. Still, water from the formations above would slip down around the casing and accumulate in the producing zone. To prevent that, part of the completion process until about 1865 included installing a seedbag around the bottom of the casing. A triumph of barnyard ingenuity, the seedbag was a leather boot, filled with flaxseed, that would expand as water infiltrated it. The combination created a seal between the wellbore and the bottom of the casing. A seedbag apparently didn't last indefinitely, so producers were pleased when J. R. Cross introduced, in 1864, an expandable rubber packer to do the same thing. Solomon Dresser came up with what was apparently an even better design for a cylindrical packer in 1880, because it launched the company that bore his name for over a hundred years until Halliburton absorbed it. (Erle Halliburton didn't found his temporarily named New Method Oil Well Cementing Company until 1919, about the same time George and Herman Brown and their brother-in-law, Dan Root, formed their own enterprise, later to be absorbed by Halliburton.)

Who's on First?

Most Canadian petro-historians become distressed whenever the name Drake is mentioned. They are quick to point out that James Miller Williams found oil in Oil Springs, Ontario, a year before Drake drilled his well. Americans are just as quick to retort that Williams only stumbled onto the oil while drilling a water well while Drake was purposefully looking for oil.

Speak to a group of Russians and they will point to an oil well in Baku, hand-dug three centuries before Drake or Williams were born.

It was not until 1903 that someone figured out that cementing casing to the borehole would eliminate almost all the water intrusion. Frustrated with leakage from the unconsolidated sands of his Lompoc, California, wells, Frank F. Hill of Union Oil Company dumped 20 sacks of cement mixed with water in the hole. He raised the casing 30 feet, capped the top, and lowered the string back to the bottom. Air pressure forced most of the cement up the outside of the casing. He still had to drill out the cement inside the casing. Eventually, he tried pumping cement down some tubing with a packer near the bottom. That eliminated most of the redrilling and began the era of modern cement jobs.

R. C. Baker, an independent oilman, saw an opportunity in 1912 and began manufacturing tools to cement casing, securing his own immortality (so far) when he named the company after himself, Baker Oil Tools.

The thought of running casing all the way through the sands containing the oil must have seemed preposterous to early oilmen (fig. 1–3), until they realized that those layers, like most others, could collapse. By 1910, blank casing was being run through the oil-bearing formations and mechanically perforated by can-opener–type blades. Pre-perforated casing, some with screens to keep out sand, became popular until about 1930. Almost inevitably, a pair of clever oilmen, Walter T. Wells and Wilbur E. Lane, constructed a gun that could shoot holes in the casing. Their first perforation job came in 1932 in a well that Union Oil thought was about played out. Wells and Lane shot 87 holes in the casing at various levels. The well came back to life and produced for several more years.

Fig. 1–3. Early drillers, ca. 1910. The floor of the drilling rig (and the crew) don't look remarkable different a hundred years later. Courtesy of Canadian Petroleum Interpretive Centre.

More Fireworks

It is amusing now to look through the eyes of the oil pioneers at notions of underground oil accumulations. The field of petroleum geology didn't exist in the 1860s. With the evidence they had at the time, some would-be geologists theorized that oil accumulated in underground fissures created by uplifts and shifts. Experience with coal-mining seams probably influenced them. Even Uncle Billy reported that his cable tool dropped six inches as he reached the infamous 69-foot depth of the Drake well.

Buying that premise and thinking about the likelihood of drilling right into one or more of these fissures, Colonel E. A. L. Roberts decided brute force might enhance the productivity of his oil wells. His Civil War experiences no doubt influenced his approach. In 1865, he lowered a flask of gunpowder into a Pennsylvania well producing three to four barrels per day and ignited it with a percussion cap. The explosion of his *torpedo*, as it was called at the time, was followed by a tenfold increase in production rate. At his next trial, two wells went from 3 each to 80 and 180 barrels per day. Roberts rushed to the U.S. Patent Office and received a broad patent on the procedure (fig. 1–4). For decades thereafter, however, lore has it that midnight explosions around the Pennsylvania countryside gave evidence of entrepreneurial patent infringement and a source of personal and litigatory frustration for Roberts.

While the geologic rational for using torpedoes proved totally misguided, Roberts spawned the notion of well *stimulation*. Still, it took more than 80 years before the Halliburton Oil Well Cementing Company offered the first hydraulic fracturing service. They pumped sand-laden napalm into an oil reservoir with enough pressure to crack open rocks, creating new channels for fluid flow. When the pressure was released, the flow of crude oil flushed out the napalm, but the sand remained behind to prop open the channels.

Only One B in Barrel

Why is the abbreviation for barrel bbl? Some time after the standardized barrel was set at 42 gallons in the 1860s, the oil industry scrambled to find containers that size. Standard Oil Company began manufacturing barrels to that specification and painted them blue to identify them. Transactions referred to the oil as coming in blue barrels, or bbls.

Fig. 1–4. Drillers lowering a torpedo into a well in 1922. Drillers to the left hold another torpedo. The pipe to the right handles the flow immediately into a tank. Courtesy of Southwest Collection, Texas Tech University.

Chemical engineers surprised petroleum engineering circles when they stumbled onto a breakthrough in well stimulation. In a happy coincidence, Dow Chemical of Midland, Texas, had an arrangement with Pure Oil Company to dispose of their voluminous quantities of waste acid by pumping them down Pure's nearly abandoned oil wells. They noticed that the more hydrochloric acid they injected into the limestone producing formation, the faster it pumped oil. After stumbling across this *Aha*! they *acidized* their first producing well in 1932, Pure's Fox #6, in Midland County. They pumped two 500-gallon slugs of hydrochloric acid and, to inhibit corrosion of the tubulars, some arsenic acid. Most agreeably, production increased fourfold. That was an improvement from 4 to still only 16 barrels per day, but they established the principle.

Underground Mysteries

Still, that was 70 years into the oil age. At the beginning, geologists had a series of doomed theories about petroleum and reservoirs. With almost total disregard for geology, the Pennsylvania oil pioneers in the 1860s punched holes with their cable tools at places that looked just like Drake's well site—in the valleys, along the creeks, near oil seeps. Few tried farther up the hills.

In 1865, oilman Cyrus D. Angell mulled over the characteristics of his few oil wells. The layering and thickness of the formations, the location of the wells, and the quality of the oil led him to his now-notorious belt theory: Oil lies in more or less continuous belts that run in a straight line from northeast to southwest; reservoirs slope away from that center line; the upper surface of a belt is level, and water courses (surface streams and rivers) are irrelevant.

Angell's unrelenting exploration successes over the next 10 years, from western Pennsylvania to southern Ohio, made it hard for geologists to argue against the theory, especially since, at the time, many of them were considered the snake oil salesmen of the oil industry. Still, the cadre at the Pennsylvania Geological Survey and the U.S. Geological Survey offices plugged away at developing an acceptable description of subsurface petroleum reservoirs. Professor Ebenezer B. Andrews almost had it right when he asserted that oil, gas, and water get trapped in natural cavities caused by uplifting geological forces that form anticlines, layers of rock pushed upward like a hump in a rug. For this, his colleagues and successors, who forgave him his obsession with fissures, named him the father of anticlinal theory.

By 1887, the notion of *stratigraphy* (the study of the history, composition, distribution, and relationships of subsurface rock layers) had captured the attention of most serious thinkers, led by the work of I. C. White, John Carll, and Edward Horton. Their report laid out the importance of porous rock layers as receptacles for oil and gas. Thereafter, stratigraphy became increasingly acknowledged as a key to describing petroleum reservoirs. So much so that John Carll labored and lobbied hard in the Pennsylvania association meetings to get drillers to note what came out as they *made hole*. In 1877, he decried, "800 miles of borehole [in the past year]…there has not been preserved a special record in one of a hundred…proper records for geologic study."

Soufflés and Sandstone

In the early 19th century, the great chefs of Dijon, France, moaned incessantly about the filthy water coming from the city mains. In 1850, Henri Darcy, the City Water Engineer, installed sand filters to purify the system. Along the way, he experimented with fluid flow through porous materials and developed equations to describe it.

For this effort, Darcy earned his own two instances of immortality: the first, the universally used darcy, a unit measuring how easily fluid flows through porous media; the second, the statue of him in place Darci, erected as an expression of gratitude by the chefs of Dijon.

Eventually, the whining and cajoling of Carll and his colleagues earned them the respect they deserved. Drillers began saving their cuttings. Sorting through buckets of rock chips was slow and tedious—which no doubt inhibited rapid accumulation of subsurface data—but the information allowed the beginnings of subsurface mapping.

Besides the boon to stratigraphic knowledge, close examination of cuttings allowed geologists to begin having serious conversations about porosity and permeability, two fundamentals of petroleum engineering. By 1880, the notion of fissures was almost totally debunked. Enough oilmen had used the fissure finder—a device that "finger felt" the sides of the borehole to locate crevices—to make them skeptical. Laboratory examination of drilling cuttings showed that oil and gas could reside in allegedly solid rock. With a few calculations using Darcy's equations (see the "Soufflés and Sandstone" box), John Carll showed that even gushers could originate in fissure-free oil sands. An emerging general theory needed only a few more pieces.

Water, Water, Everywhere

To the early operators, water was more than a nuisance. It was a menace. In the beginning, oil-field production philosophy most often called for drilling up a field and producing at maximum natural flow rates for as long as possible. After that, pumps were used in the wells until the amount of water coming up with the oil made pumping uneconomic. Water was a menace.

The Lufkin Pump

In 1923, Ross Sterling, the president of Humble Oil and Refining Company and a future Texas governor, invited Walter W. Trout to lunch to discuss a vexing oil field problem—finding a reliable pumping mechanism for low-pressure oil wells. Trout returned to his modest machine shop and proceeded with a few unsuccessful prototypes. At another lunch, this time with W. L. Todd, of Standard Oil, Trout showed a sketch of a pumping unit built around a counterbalance principle. Todd liked the idea and had Trout's company, Lufkin Foundry and Machine Company, build and install them in his Hull, Texas, field. Trout patented the design in 1925, and the Lufkin pump jack design became an enduring world standard (fig. 1–5).

Fig. 1–5. An early design of a Lufkin pump jack. Today's pump jacks look muck like they did 60 years ago. Courtesy of Lufkin Industries.

Evidence and contradictory notions started to emerge. In 1865, the operators of the prolific Bradford Field in Pennsylvania noticed an increase in oil production after some surface water inadvertently poured down some adjacent, dormant oil wells. Inspired by this observation, the operators pumped water into the edge wells of the nearby Pithole Field and watched as the flow from their oil producers steadily rose. *Waterflooding* had begun.

About that time, geologists started to speak about the idea that oil generally sat on a layer of water, especially in anticlinal reservoirs. Some others vehemently disagreed. In fact, the State of Pennsylvania summarily prohibited pumping water into an oil field and didn't reverse the regulation until 1921. By that time, overwhelming evidence contradicted even the most stubborn bureaucrat. Another brick in the foundation of reservoir engineering had found its place—water replacement as a source of energy that could push oil to the surface.

Breathings of the Earth

Like water, for a long while natural gas was at least a nuisance and sometimes also a menace. Certainly, minor amounts of gas found their use at the well site as boiler fuel for steam engines powering the pumps. A limited market for natural gas in nearby cities for heat, power, and light absorbed some. But mostly, oilmen flared their gas throughout the 19th and well into the 20th century just to get rid of it.

Only 2 of the first 125 wells in Pennsylvania flowed unassisted. The rest were pumpers. Then came a series of gushers: 400–500 barrels per day. That started geologists thinking about the reasons. J. P. Leslie, another insightful member of the Pennsylvania Geological Survey, put the pieces together in 1865. He wrote that dissolved gas and water have enough energy in some reservoirs to force oil out of the formation and up the wellbore for long periods of time. The crude oil and natural gas would arrive at the surface "just like seltzer from a soda fountain."

John Carll, his colleague, had a headier analogy: Drawing oil from the rock may be compared to drawing beer from a barrel. The barrel is placed in the cellar and a pump inserted. At first the liquor flows freely through the tube without using the pump, but presently the gas weakens and the pump is called into recognition. And finally, the gas pressure becomes so weak that a vent hole must be made to admit atmospheric pressure before the barrel can be emptied, even by the pump.

Obviously, Carll spent more time in pubs than Leslie, but his further insights contributed the elements of reservoir energy and the forces of water and natural gas to the general theory. For this and many other contributions, industry sages nowadays consider Carll the Father of Reservoir Engineering.

Crisis and Reservoir Engineering

By the 1880s, operators began to buy into the idea of conserving reservoir energy by putting a choke on the producing lines to hold pressure on the reservoir. But conservation of natural gas was not yet the objective. To the contrary, billions of cubic feet of gas were flared through most of the 20th century in the United States (fig. 1–6), as well as in the Middle East and western Africa. The markets for gas were just too remote from those oil wells that contained associated gas. Automobile drivers could turn off their headlights when they passed many oil fields at night. Gas flaring turned night into day. Understandably, the waste bothered many conservationists and environmentalists, who were not necessarily the same people at that time. The conservationists worried about the economic recovery of the resource in the ground; the environmentalists were tormented by the impact of wanton production on the ecology.

The advance of technology sometimes had contradictory effects. One of the hazards cable tool drillers always faced was unexpectedly tapping into a high-pressure reservoir. Nothing blocked the space between them and the fluids that came gushing up the borehole. The walls of museums in the oil patch are replete with nostalgic pictures of blowouts—hydrocarbons spewing a hundred feet into the air.

From *Derrick's Handbook of Petroleum,* **1898:**
The Armstrong 2 well in Butler County, Pa., was thought to be a duster, until torpedoed. One excited observer wrote, "the grandest scene ever witnessed in oildom. When the shot took effect, barren rock, as if smitten by the rod of Moses, poured forth its torrent of oil. It was such a magnificent and awful a spectacle that only a painter's brush or poet's pencil would do it justice... the loosening of a thunderbolt. For a moment the cloud of gas hid the derrick from sight, and then as it cleared away, a solid gold column... shot from the derrick 80 feet through the air."

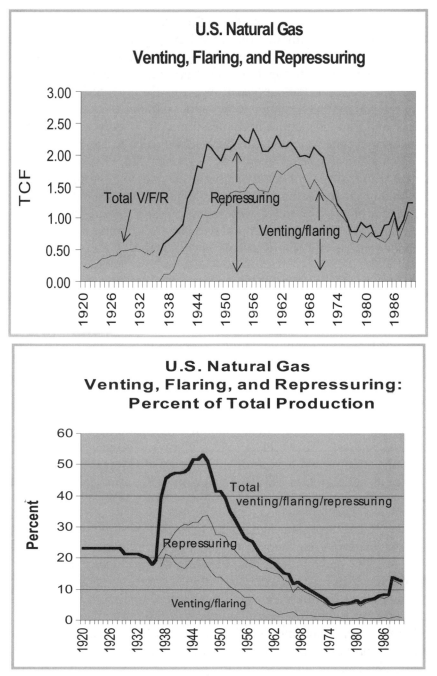

Fig. 1–6. Natural gas flaring in the United States. Trillions of cubic feet of gas have been vented or flared. As late as the mid-twentieth century, 20 percent of the produced gas was being wasted. But by the last quarter century, nearly all produced gas was conserved.

The introduction of *rotary drilling* at Spindletop in 1901 was a giant step toward controlling many of these situations (fig. 1–7). Part of the rotary drilling design called for circulating *drilling mud* (a slurry of water and clay) down the borehole with the drill bit, with providentially beneficial results. Not only did the mud remove the cuttings from the hole, cool the drilling bit, and keep water from seeping into the upper parts of the hole, but it added weight and enough bottom-hole pressure to help prevent hydrocarbons from escaping the hole uncontrollably. (Ironically, at the first attempt at Spindletop, the hole blew out anyway.) Around 1926, Jim Abercrombie, an oilman who had nearly been killed in a recent blowout, brought his phobia to a young machinist named Harry Cameron. Together (fig. 1–8) they designed a blowout preventer whose basic design oilmen still use today. With these two innovations, countless volumes of oil and gas have been contained that would otherwise have been vented to the environment.

Yet rotary drilling continued the inexorable march of efficiency that let oilmen reach deeper, more hidden, and more nearly unreachable targets and bring increased volumes of oil to market.

It was overproduction of oil that triggered public policy remedies. As the giant oil fields of Texas and Oklahoma came onstream at the beginning of the 20th century, a stampede to produce every oil field as rapidly as possible (fig. 1–9) laid waste to the economics of the industry. Prompted by the court-established *rule of capture,* any landowner could—and usually did—drill a well to tap the petroleum that lay beneath. On the 92,000-acre East Texas Field, 3,612 wells were drilled. Wasted oil ran down the streams and bayous of the Southeast. Oversupply drove prices to 10 cents per barrel. In 1931, the governor of Texas declared martial law to control the chaos.

Henry L. Doherty, an oilman and firebrand of some repute, captured the imagination and support of the conservationists. Together, they badgered the governments of the largest producing states—Texas, Oklahoma, Kansas, Colorado, and Illinois—into forming the Interstate Oil Compact Commission (IOCC). Initially, the IOCC instituted voluntary cutbacks of oil production, but in 1935, mandatory reductions, called *proration*, were introduced. The Texas Railroad Commission and its counterpart agencies in the other states set and enforced the rules. The federal government supported them by passing the Connelly Hot Oil Act, which prohibited moving excess oil production across state lines. It also established the authority of the state commissions to regulate well spacing, limit individual well production rates to protect reservoir pressures, and encourage unitization.

Fig. 1–7. The first rotary drilling job. Rotary drilling came to the U.S. first at Spindle top in 1901. Courtesy of Spindletop Museum.

Fig. 1–8. Designing a blowout preventer. Jim Abercrombie and Harry Cameron drew the first BOP in the sand on the floor of Cameron's machine shop, ca. 1926. Courtesy of Cooper Cameron.

Everyone got something at the stroke of a pen: proration meant higher, stable prices for the producers; conservationists and environmentalists were mollified; the state bureaucracies had a legitimate empire to administer; and the science of petroleum engineering gained new prominence.

To prorate, every well needed a documented, authorized *maximum efficient rate* (MER) of production. Engineers had to take into account the reservoir pressure, porosity, permeability, gas/oil ratio, water cut, and more at various production rates. They wove this into a story about

Fig. 1–9. Spindletop, a year later, 1902. Courtesy of Spindletop Museum.

ultimate recovery and calculated an MER. Proration below 100% (roughly the sum of all MERs in an area or state) lasted until 1970 in Texas, but state regulations still require reservoir engineers to apply for a permit to establish the MER of many types of wells.

Horizontal Drilling

Drilling vertical wells into horizontal reservoirs bothered some production engineers for a long time. Almost all reservoirs are wider than they are deep, so a vertical wellbore contacts the hydrocarbon interface in the wrong plane. Elf Aquitaine finally exploited some enabling technologies that let them do the precursor to *horizontal drilling*. In 1980–83, they successfully penetrated several reservoirs in France and Italy with nearly horizontal wells and satisfactorily improved their productivity. (The first horizontal wells recorded were drilled in Texon, Texas, in 1929, and Venango County, Pennsylvania, in 1944, but sadly, these efforts generated little commercial interest at the time.)

To accomplish the turn from vertical to horizontal, the drillers could not rotate the drill pipe without buckling it. Instead, they used a downhole motor, which turned only the drill bit. They began a wide turn, with a radius of 1,000–2,500 feet, steering the drill bit to the target, using downhole devices that diverted the drill bit in small increments and downhole telemetry to track the course. Still, although the technique has matured, horizontal drilling is far more painstaking than conventional drilling; however, the extra expense has often proven rewarding.

A Golden Decade

In the 1930s, innovation and consolidation of intellectual and practical knowledge permeated the upstream industry in a production renaissance. During this time, companies routinely adopted

- drilling rig instrumentation and mud control
- side-wall coring
- bullet perforation
- squeeze cementing
- internal combustion engine power to replace steam
- downhole pumps

Over the same decade, geologists and engineers regularly used

- calculation of porosity and permeability
- downhole pressure and temperature measurements
- fluid flow and material balance equations, based on reservoirs as common containers of both hydrocarbons and energy
- reinjection of water
- acidizing
- drill-stem testing
- seismic mapping
- electric and radioactive logging

What all these are is what this book is about.

The Great Offshore

Until well into the 20th century, oilmen had to start geologic surveys with surface observations. That limited their efforts to the onshore. Even the first offshore production—at Summerland, California, in 1897—was just an extension of onshore fields. The operator built piers into the Pacific Ocean to follow the reservoirs beyond the shoreline (fig. 1–10).

Operators coped with exploration and production operations in the shallow inland waters like Lake Caddo, Louisiana, starting in 1908 and the more prolific Lake Maracaibo, in Venezuela, starting in 1924 by building platforms on top of timber or concrete pilings. Even at these locations, geologists sited their wildcats by observing oil and gas seeps.

Fig. 1–10. Piers and derricks at Summerland, California Oil, 1901. Courtesy of USGS.

It was the advent of seismic technology that enabled offshore exploration by allowing a look at the subsurface without surveying the surface. Kerr-McGee grabbed the brass ring for the first successful wildcat and production at a location out of sight of land. In 1947, they placed a small and largely prefabricated platform on a site their seismic surveyors recommended, in the Ship Shoal area, 40 miles from the Louisiana coast. The platform rested on steel pilings, linked together like Tinker Toys, to withstand the occasional hurricane-force winds and waves. Flushed with exploration success, they used the same platform for their production operations.

Later, offshore exploration switched to the more mobile and less expensive submersible, semisubmersible, and drillships, but for almost the balance of the 20th century, permanent offshore production facilities in ever-deeper water sat on top of ever-taller steel platforms. Shell Oil installed the tallest conventional steel production platform, Bullwinkle, in 1,326 feet of Gulf of Mexico water in 1989. The supporting structure needed 44,500 tons of steel and another 9,500 tons of piling to anchor it to the bottom.

Bullwinkle became a watershed. The increasing number of prolific reservoirs in the deepwater (depths of more than 1,500 feet) could not afford the cost of conventional platforms. In the 1980s, Petrobras had already chosen an alternative approach to develop a continuing string of exploration successes in the very deep waters of the Campos Basin of offshore Brazil. They placed the wellheads of their producing units on the seafloor and produced up risers into floating production facilities.

By extension, they were able to develop the giant Marlem Field in 3,369 feet and the Marlem Sul Field in 5,666 feet at a rapid pace in the 1990s (fig. 1–11).

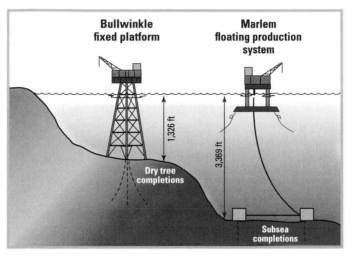

Fig. 1–11. Marlem and Bullwinkle. Two different approaches to production platforms as industry ventured into deeper water.

In the Gulf of Mexico, operators initially chose tension leg platforms and lighter and more flexible compliant towers. From both, they could drill their development wells and produce, even as the platform swayed around the drill site thousands of feet below. By the end of the 20th century, the list of offshore development systems being used around the world also included *spars* (huge, buoylike storage containers) and *FPSOs* (floating production, storage, and off-loading vessels), especially in waters deeper than 1,500 feet.

From its real beginning in 1947, offshore development has been a continuous battle against a ticking clock. At any instance, the cost of almost every task during the development of an offshore discovery escalates almost exponentially with water depth. But, driven by an imperative to improve economics, advances in tools and procedures have continuously lowered the cost curves. And three factors have enabled an inexorable march into deeper waters:

- geologic discoveries of accommodatingly large accumulations of hydrocarbons—reservoirs routinely of hundreds of (instead of tens of) millions of barrels

19

- reservoir conditions that allow prolific, world-class wells—production in the thousands of barrels per day (instead of tens or hundreds)

- persistent technological progress that continues to enable access to ever more absurdly located reservoirs under impossibly difficult sea conditions

The end of history is not in sight. That becomes promisingly evident to the most casual observer of the pages of the oil and gas industry journals, the trade shows, and the expositions, all conspicuously rich in innovation and exuberant optimism.

2

The Container:
The Reservoir

Don't throw the old bucket away until you see if the new one holds water.

—Swedish proverb

Producing oil and gas has a lot in common with drinking a soda. After all, what do people do when they get thirsty as they drive down the road? They look for a convenience store (explore), choose a cup or a bottle (the reservoir), stick in a straw (drill), and suck out the refreshment (produce it). While thirsty people might want to know a little about the container—cup or bottle, ice or not—they probably don't care much about its origins. But petroleum engineers might, and surely they want to know a lot about the container besides the oil and gas that is in it. This chapter flows through four parts:

Origin   Transformation   Characterization   Discovery

Deep background—*really* deep—may not help anyone boost the production from a well, but it's intriguing to think that the story about petroleum coming out of the ground has its origins in the Big Bang theory. After reviewing that backdrop, the story gets closer to the oil business by looking at the transforming forces that created petroleum reservoirs. Finally, the discussion arrives in the midst of something production people are always curious about, what reservoirs look like and how they are found.

Origin

ORIGIN   Transformation   Characterization   Discovery

Physicists measure the diameter of the Earth at the equator to be 3,961 miles, but petroleum geologists are only interested in the outer five to seven miles. This thin layer comprises less than 0.2% of the Earth's volume, yet almost all earth scientists think it contains all the petroleum that can ever be produced. All young geologists today begin their understanding of this thin skin by learning about its origin and the forces that acted on it.

Cosmologists have studied the speed and direction of the zillions of celestial bodies in the universe—no small task for the few hundred people in the world who do this. Over the past few decades, they have concluded that the universe formed at a central point about 14 billion years ago with a cosmic explosion of unimaginably huge force, the Big Bang. Since then, every particle within the universe has been moving away from that central point, continuously expanding the size of the universe.

The Big Bang created primordial clouds of gases, mostly simple molecules like hydrogen, which later concentrated into hot fireballs (stars), or so the theory goes. Some hydrogen morphed to helium over a few million

---

The Cosmic Hierarchy

THE UNIVERSE—everything, all matter as we know it.

A GALAXY—a system of stars, planets, black holes, gases, and cosmic dust particles. (The Milky Way is the galaxy in which the Earth's solar system exists. Some say it contains 100 billion stars. They also say that there are 100 billion galaxies in the universe.)

A SOLAR SYSTEM—a sun surrounded by its planets and their moons, with meteors and other miscellaneous cosmic matter.

A SUN—a star. Every star has its own measure of heat, size, age, and composition. Some stars have planets orbiting them.

A PLANET—a concentration of material orbiting a sun. Our sun has nine planets in orbit. Earth is third closest to our sun.

A MOON—a concentration of matter orbiting a planet. Earth has 1, Jupiter has 16, and Mars has 2.

METEORS and ASTEROIDS—miscellaneous planetlike fragments that have various paths through a solar system.

years and eventually into heavier molecules like carbon, nitrogen, iron, and nickel. Stars collapsed on themselves in monumental events, the creation of supernovas, which blew the stars apart. From that, over a few million more years, gas and cosmic dust condensed again, forming a new star in a swirl of dust and gas. The forces of gravity shaped this mass of stellar material into disklike systems, where the heavier elements formed planets orbiting a hydrogen-based sun. In the case of the Earth's solar system, the planets had rocky cores surrounded by gaseous atmospheres—some thick, others thin.

Down to Earth

The Earth is one of the smaller planets of this solar system. In its youth, it was a mass of dust, subatomic particles, and gaseous material all connected under the forces of gravity. As it condensed into a solid mass, material rich in iron, nickel, and other dense minerals swirling in the solar system bombarded it. The energy released by this meteoric bombardment kept the primitive Earth in a molten state. The forces of gravity pulled the heavier material to the center; lighter minerals floated to the surface. As the meteoric impacts abated, the Earth cooled, and the lighter material solidified and became the crust.

J. R. R. Tolkien may have called it the *mittel earth,* but geologists say that at the center of this planet lies the *core,* which has two parts, a solid, very dense center surrounded by a liquid shell—more or less the opposite of a bagel. Scientists have determined its size, density, mobility (plasticity), and composition by tracking seismic waves generated by earthquakes. Seismic receivers sited around the Earth detect the speed and deflection of these seismic energy waves as they pass through and around the core.

Above the core and below the crust is a zone of plasticlike material known as the mantle (fig. 2–1). This zone is kept at elevated temperatures by radioactivity. Within the mantle, giant heat cells or convection currents push the mantle material around in almost dismissively small flows. Over the eons, these convection currents forced molten rock through breaches in the Earth's crust—rifts and volcanoes. Gases spewed out and became part of the Earth's atmosphere. Condensation of the gases produced water that rained on the protruding crust and formed oceans. These rains plus waves washing against the landmass eroded the crust and, through the action of streams and rivers, deposited sediments that became rock layers. This building and wearing down of the landmass repeated itself over hundreds of millions of years. Some of the oldest rocks on the Earth's surface, nearly four billion years old, are found in northeastern Canada. They are remnants of these early lands.

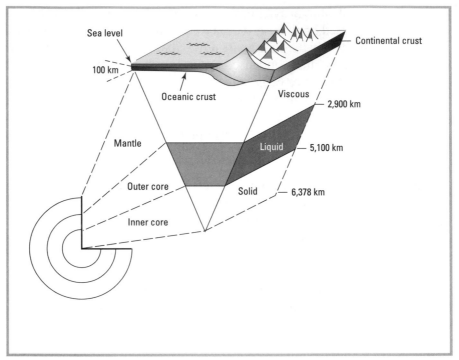

Fig. 2–1. Partial section of Earth

Moho

Only a hundred years ago did even the best scientific minds start to understand the make-up of planet Earth. At the time, most knew that if they dug a hole, eventually they would encounter rock, but beyond that was a mystery.

Then, in 1906, an Irish geologist, R. D. Oldham, improbably studying the seismic record of a Guatemalan earthquake, noticed that the sound waves had bounced off a formation deep within the Earth. With a burst of insight, he concluded that the Earth had some kind of core.

About the same time, a Croatian seismologist, Andrija Mohorovičić was analyzing a Zagreb earthquake and stumbled across a much shallower but more pronounced deflection. He had discovered the boundary between the Earth's crust and mantle. Since that time, that margin has been known as the Mohorovičić discontinuity, or to elocution-challenged geologists, the Moho.

The crust, the mantle, and the core had been identified.

Pangaea and the shifting plates

Most scientists buy the modern theory that about 200 million years ago what are now our modern continents were all part of a single mammoth continental body, now called *Pangaea* (fig. 2–2). (The name "Pangaea" comes from a Greek word meaning "the whole Earth.") Millions of years after the formation of this continental body, rising molten rock sundered this large landmass into numerous plates. Since then, the plates have moved around on the surface of the Earth, floating like lily pads on a pond, in a phenomenon called "plate tectonics" (fig. 2–3).

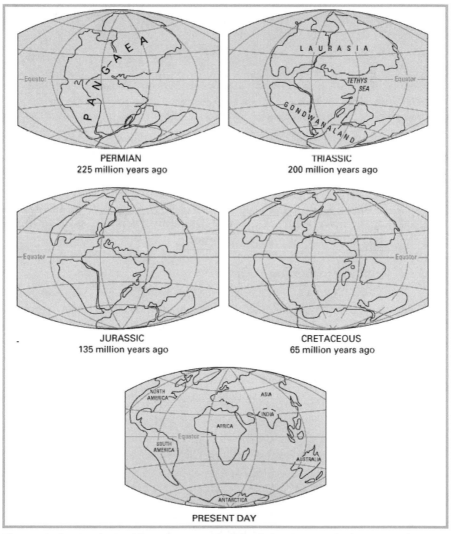

Fig. 2–2. Reshaping the Earth's continents. A single landmass, Pangaea, broke apart and moved to create today's continents. Courtesy of USGS.

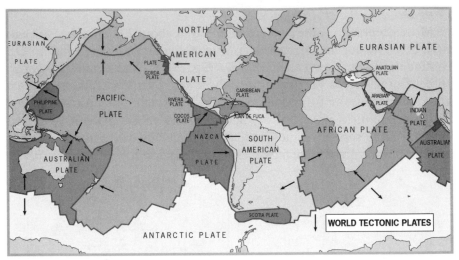

Fig. 2–3. The world's tectonic plates. Oceanic and continental plates are moving in the direction indicated by the arrows. Courtesy of USGS.

The similarities in the coastlines of South America and Africa have long attracted the attention of geographers and geologists. Even to a child's eye, they fit together like pieces of a jigsaw puzzle. The theory that continents drifted on the Earth's crust was proposed in the early 20th century, generating a disparaging clamor from the scientific community. The critics asserted that the laws of physics would not allow the landmasses to move over the rigid material making up the mantle.

Thinking in Geologic Time

In 1908, an amateur American geologist, Frank B. Taylor, proposed a theory that the Earth's continents had drifted around the globe, a theory that accounted for the cozy, corresponding shapes of some—most notably Africa and South America. Taylor was at first ignored then reviled and ostracized by the profession. Even as late as 1965, the *Journal of Geophysical Research* refused articles that reaffirmed Taylor's theory.

During World War II, the U.S. Navy identified a lengthy discontinuity in the Atlantic Ocean floor. Building on that work, Harry Hess at Princeton and later Drummond Matthews at Cambridge described the Atlantic Rift, a 12,000-mile, north-south crack in the Earth's crust. They clearly demonstrated that seafloor spreading was an identifiable source of continental plate movement.

Geophysicists perhaps ponder in geologic time because it wasn't until the 1970s that most of them accepted plate tectonics as valid.

Generally accepted theory now has the Earth's crust made up of two types of plates, *oceanic* and *continental*, although neither contains exclusively oceans or continents. The oceanic plates are relatively thin (only five miles), composed of basaltic material (dense, dark rock with less than 52% *silica* [the element *silicon* plus oxygen, SiO_2]) and mostly covered with water. The continental plates are thicker (ranging up to 30 miles) and composed of lighter, granitic igneous rocks (light colored, coarse grained, and having more than 65% silica) and sedimentary sequences.

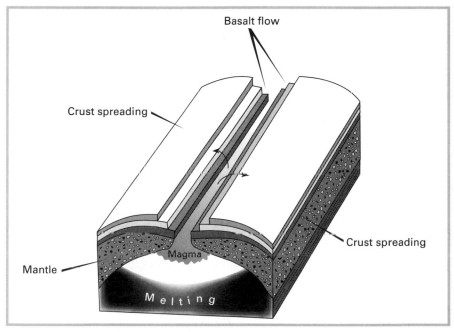

Fig. 2–4. Seafloor spreading. Oceanic plates are forced apart and the resulting rift is filled with molten rock from the mantle.

According to the *seafloor spreading theory*, convection currents in the mantle crack open the crust, force it apart, and push new plate material up through thcsc rifts (faults) in the seafloor (fig. 2–4). As the seafloor spreads, the ocean plates slowly crunch against continental plates with great force. The oceanic plates slip under the continental plates, in a process called *subduction*, causing great stress and forming both mountains and subsurface distortions (fig. 2–5). Continental plates, forced by convection currents within the Earth, continue today to move away from and back against one another. Continental plates and the basins at their margins that have been collecting the sediments resulting from erosion are continuously contorted by the impact of the plates. In those places, accumulated layers of sediment from erosion can be thrust upward.

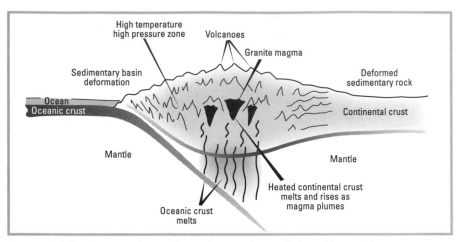

Fig. 2–5. Subduction. One plate pushes against another and is forced beneath it.

Within the plates, away from the margins, compensating forces can cause ripping and tearing of the continental landmasses. The two continental plates that form the Indian subcontinent and Eurasia are moving against each other at the rate of two inches per year. The Himalayas were created in the process. On the other side of the globe, the Andes were formed and are still being shaped by an oceanic plate (the Nazca plate) moving against a continental plate (the South American plate). Where the plates move sideways against each other, as along the San Andreas Fault and offshore Sumatra, the collisions create devastating earthquakes and sometimes tsunamis.

Not all continental margins are *active* like the ones that formed the Coast Ranges of North America and the Andes. The eastern edges of the North and South American plates are *passive*, with no bumping or grinding between the plates. To the contrary, the African and European plates are moving away from the American plates by seafloor spreading.

Transformation

Origin ▸ TRANSFORMATION ▸ Characterization ▸ Discovery

Despite the Herculean forces that go into creating them, mountains are not permanent—they are worn down over time by *erosion*. The material (sediment) from this erosion is washed down and deposited in layers, some remaining in riverbeds, some landing in lakes, and some in the sea.

Millions of years later, what was subsea might then be on land, and vice versa, as plate tectonics continues to distort the continental plates and shift shorelines back and forth.

The rock that lies above sea level is continually buffeted by condensing water vapor (rain) and by the action of the waves of the surrounding oceans. These forces erode the surface; rock fragments, large and small, move down mountainsides, across the land in rivers, and out into the surrounding lakes and seas. This eroded rock material eventually drops out of the conveying currents to the lakes and seafloor, forming ever-thickening deposits of granular material. (Illustrations of all this come up later, in figs. 2–8 and 2–9.)

Not all oceans surrounding the continental masses have been invaded by rivers carrying eroded rock chips from surrounding highlands. In some, chemical reactions caused a different kind of sediment to form. In cooler regions of the oceans, chemical precipitation occurred, forming lime oozes, slimy-sounding fine matter consisting mainly of calcium carbonate ($CaCO_3$) plus other minerals, as well as the shells from myriads of once-floating minute organisms. In warmer seas, larger forms of shell life drew calcium carbonate from the circulating water to form coral reefs.

So, both erosion and deposition within the ancient oceans of the planet produced a variety of deposits, a process that starts to get the interest of petroleum geologists. Where there was little influx of rock material, chemical deposits, mostly lime oozes and coral reefs, predominated. Where the influx was more solid material, deposits of sands, silts, and clays accumulated.

Shorelines have not remained constant over geologic time. Near-shore deposits (the coarser material) have accumulated continuously as the shoreline has moved back one way and then reversed itself. Simultaneously with the deposit of sand near shore, silts have been deposited farther offshore and limes even farther away. As the position of the shorelines has changed, so have the relative positions of the sedimentary sequences. The results have been sheetlike layers of the same type of sediment over hundreds of miles.

Successive accumulations of these materials formed layers of increasingly compacted material. Clays became shales, particle layers became silts and sands, and the chemical and organic deposits became lime muds. As these layers were buried and compacted, they became increasingly harder and more *competent* (difficult to rupture or fracture).

How many millions of years did all this take? Radioactive dating has shown some sandstone in Australia to be 3.8 billion years old. Mineral grains (zircon) in that sandstone have been dated at 4.5 billion years, so the Earth is at least that old. When the first consolidation of the particles from the Big Bang occurred remains a matter of conjecture. The question motivates scientists to expend large efforts on projects using the Hubble

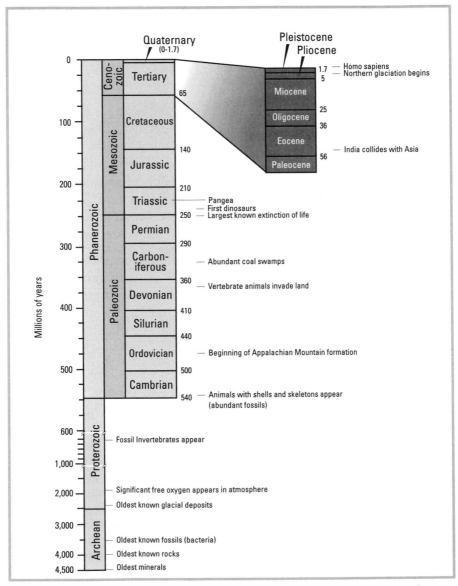

Fig. 2–6. Geologic timescale. Times are shown in millions of years before the present. A few important events in the Earth's history are indicated.

telescope and other instruments to gaze further into outer space, but current theory places the age of the universe at around 14.5 billion years.

Over the past 200 years, geologists studied the similarities of rock layers and realized that younger rocks lie above older rocks. More recently, they used radioactive dating and examination of tons of fossils recovered from various layers to develop the geologic timescale in figure 2–6. They concluded that deposition of sediments forming the layers did not occur simultaneously around the world. Where there is an absence of sediments (a break in the continuous layering) at a particular geographic spot, there is a skip, known as an *unconformity*, in the sedimentary sequence. Unconformities come in various forms (fig. 2–7), but invariably they capture the attention of petroleum geologists because they are an important element in trapping oil beneath the surface. (More on unconformities follows later.)

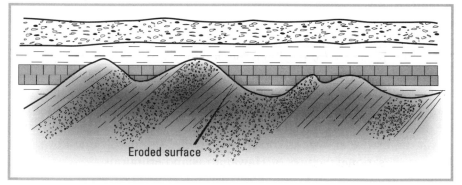

Fig. 2–7. An unconformity. In this case, an angular unconformity. Rock layers have been tilted, raised above sea level, eroded, and then covered by the sea. Afterwards additional layers were deposited horizontally.

Characterization

Origin　Transformation　CHARACTERIZATION　Discovery

The building blocks of the Earth's crust—its various rock types—are minerals, and the physical and chemical character of a rock is called its *lithology*. Minerals are formed by combinations of elements and compounds. By far the most abundant are the silicates formed from silica and various metals.

Is? Was? Being?

Ask geologists and geophysicists about relevant history, and they will speak in millions and hundreds of millions of years. Understandably, they use the past tense. But geological change didn't stop when the science of geology formally organized itself 150 years ago. The forces continue unabated. They move so exasperatingly slow, however, that almost all the relevant results took place well in the past. In this book, geologic discussions usually use the past tense ("was eroded"), rather than the present ("is eroded") or the present participle ("is being eroded")—but not always.

There are three groups of rocks: *igneous, sedimentary*, and *metamorphic*. With only a few exceptions, sedimentary rocks capture the entire attention of petroleum geologists.

Rock of (various) ages

Igneous rocks are formed from the bubbling up of the molten material (magma) that forms the mantle of the Earth. The superheated magma reaches the surface through volcanic eruptions, deep fissures into the mantle, or plumes of magmatic material. The most familiar igneous rocks are *granite* and *basalt*. Granite forms much of the central Rocky Mountains. Basalt, otherwise known as cooled lava, comes from volcanic flows like those in sensational pictures of Hawaiian Island eruptions.

Sedimentary rocks comprise the bulk of the Earth's continental crust. They are of two types, *clastic* and *chemical* rocks. The clastic rocks come from the breakdown and erosion of igneous, metamorphic, and earlier formed sedimentary rocks. All of that forms the mountains and highlands of the continents.

Rapid rock erosion (an oxymoron) occurs in the higher elevations of continents when water seeps into crevasses and cracks, freezing and splitting the rock apart. Gravity, avalanches, and glaciers carry the fragments down the hills and mountains to streams and rivers that carry the rubble to the point where the transporting energy of the river can no longer support the carried material (fig. 2–8).

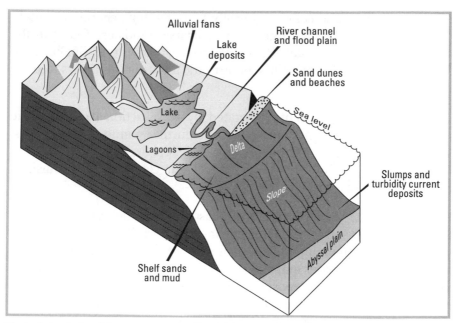

Fig. 2–8. Eroded rock fragments moving from the mountains to the sea

Clastic sedimentary rocks are named according to the size of their fragments:

- *gravels*—the largest material
- *sands* (coarse, medium, and fine grained)
- *silts*
- *clays*

While sands have a grain size distinguishable by the naked eye, silts are so fine grained that they can be distinguished from clay only through microscopic examination. Clays have no distinguishable grains whatsoever. In the field, meticulously avid geologists sometimes chew a small rock fragment and distinguish a fine-grained silt from a clay by feeling the grit of the silt between their teeth.

Going with the flow

The largest clastic material comes to rest closest to its point of origin, and the finest material moves the farthest: boulders lie closest to the mountains, and very fine, small-grained silts and clays usually find their final resting place in lakes and seas. In the continental United States, for example, boulders and cobbles typically lie in the high-energy mountain streams of the Rockies, and gravels and coarser sands are found in the upland rivers (the San Juan, Platte, Madison, Yellowstone, Snake, etc.). The Missouri and Mississippi rivers carry the finest silts and clays into the Gulf of Mexico.

Once in the sea, currents generated by wave action can move the sediments along the continent's shoreline, further winnowing the finest material from the coarser detritus. The Mississippi dumps its load into the Gulf, and the westerly currents carry the silt and clay along the Louisiana coast toward Texas.

The sedimentary cycle is repetitive. Clastic material deposited in ancient seas is compacted and consolidated into layers of rock. Those layers are later lifted by tectonic (mountain-building) forces or by the upward intrusion of masses of molten igneous material, are again eroded by weather and carried toward the seas, and are deposited once again in a mixture of sedimentary materials. The new deposits may or may not have the mineralogical and physical characteristics of the original sedimentary rock.

Where there is little or no clastic material in the oceans, shell-forming sea life abounds. These animals are principals in the formation of chemical sediments, the second major type. Corals are the dominant rock-building mechanism. These animals develop shells of calcium carbonate and live together in colonies called coral reefs. The animals eventually die, and their shells break up and gather as sediment around and throughout the growing reef. Over long periods, a *lithification* process can transform the accumulation of fragments and whole corals into the type of rock called *limestone*. One of the most impressive examples of a coral reef from an ancient sea that no longer exists is El Capitan, a mountain just north of El Paso, Texas.

Lithification

As additional clastic material is carried into the oceans and deposited on the *gravels, sands, silts,* and *clays* already there, the underlying material is compacted into conglomerate, sandstone, siltstone, and shale, respectively (fig. 2–9). In the compaction process, seawater is squeezed from the sediment, and the mineral grains finally touch and then support each other. Silica and calcium carbonate dissolved in seawater circulating through the spaces between the mineral grains precipitate around the grains. This precipitation acts as cement, creating, for instance, sandstone from an otherwise unconsolidated, loose sand.

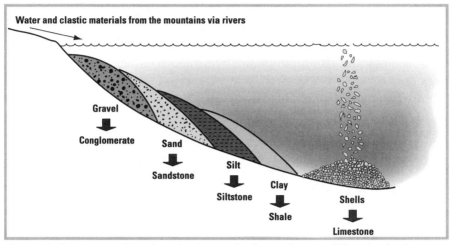

Fig. 2–9. Sediments to rock. Various size classes of sediments become different rocks after lithification.

Likewise, limestone is formed by compaction of lime mud through accumulation of successive layers of chemical and organic material. *Dolomite*, another type of chemical rock, differs from limestone in that magnesium (Mg) has combined with the calcium during circulation of the sea water.

Sodium chloride (NaCl), plain old salt, is on the list of chemical deposits important to the oil industry. Salt, as an *evaporite*, formed in sheets by the evaporation of saltwater over a long period of time. The Great Salt Lake, in Utah, and its surrounding desert formed when evaporating water that was never replaced by rain or runoff concentrated the salt content. Similarly, the huge salt deposits in the sedimentary layers of the Gulf of Mexico and the North Sea are among the largest of the ancient salt deposits. They formed during long intervals when these bodies of water became landlocked and evaporated.

Metamorphic rocks, the third major category of rocks, are sedimentary rocks that have been heated by proximity to igneous rocks or magmas or have been subjected to the high pressures and temperatures of deep burial within the Earth's crust. Slate is metamorphic rock formed from shale, marble is formed from limestone and dolomite, and quartzite is formed from sandstone. Generally, some of the oldest known rocks are metamorphic. They can be either sedimentary or igneous rock that have been deeply buried and then sometimes uplifted close to the surface, again by tectonic, mountain-building forces where they are seen today.

The Earth's history has been one of repeating cycles of deposition, mountain building, erosion, more deposition, and more mountain building. Coastlines moved back and forth as continental plates drifted and as volcanoes and rising mountains reformed continental margins. What was once a near-shore environment where sands and silts were deposited became an onshore region far from the sea or an offshore area where the

Fig. 2–10. The Grand Canyon. The Colorado River cut into (eroded) horizontal rock layers. Courtesy of James Popple

currents carried nothing but the finest silts and clay. As a consequence, sedimentary layers were laid down with varying rock types (lithologies), alternating one over the other. Familiar vistas of the Grand Canyon (fig. 2–10), of mountainsides and sea cliffs (fig. 2–11), and even of local highway cuts throughout the countryside show layers of sandstone, shale, and limestone in infinite and stunning varieties of sequences often highly contorted into remarkable folded forms.

Fig. 2–11. A sea cliff. Sedimentary layers, at one time horizontal, have been distorted into folds by tectonic forces. Courtesy of Lionel Weiss.

Spaces

Load up a soda cup with crushed ice like they do at McDonald's and then fill it with soda. Then pour the soda into another, identical cup. The soda takes up only part of the second cup and is equal to the space around the ice. If you were to substitute perfect spheres the size of marbles for the ice, the soda would amount to just 48% of the cup's volume (fig. 2–12). Sand grains are far from spherical, yet they touch and leave room in the same way. This void between them is called *pore space*, and as a percentage of the whole rock, it is known as *porosity*. Geologists think of porosity as the storage capacity of a rock.

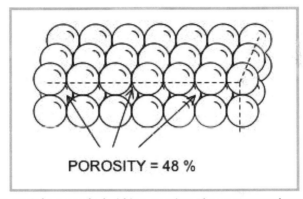

Fig. 2–12. Spheres stacked within a container. The empty space between spheres equals 48% of the total volume.

The shape of the sand grains depends mostly on the distance that they traveled. During transportation, the sand grains collide, rub, and wear away each other. Sharp, angular grains become rounded and smoothed in their travel from their origin to their point of deposition. Forensic geologists know that the more angular are the grains they find, the closer to their point of origin they lie.

The least-rounded grains settle to the ocean floor in patterns leaving less porosity than the more rounded grains (fig. 2–13). Sand deposits usually contain more rounded material than do the deposits of smaller-grained silts. As a consequence, sands and their compacted counterpart, sandstone, are more porous than are silts and siltstone. Poorly sorted sediments, those that have not been transported far enough for the forces of gravity to segregate and screen the larger grains from the smaller ones, have less porosity than do well-sorted deposits. The smaller grains actually fill the spaces between the supporting larger grains.

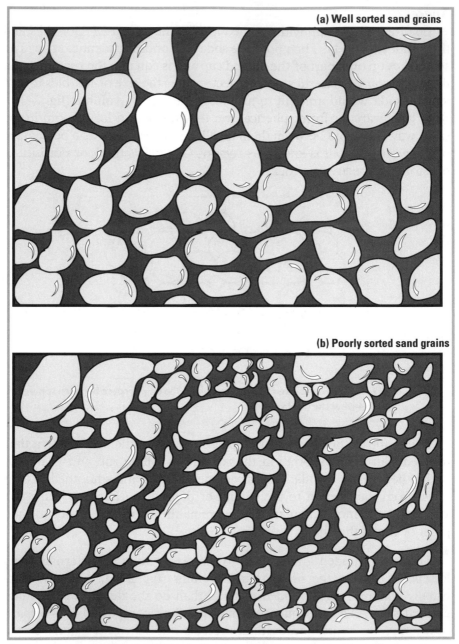

(a) Well sorted sand grains

(b) Poorly sorted sand grains

Fig. 2–13. Poorly sorted grains. A blend of large, rounded gains mixed with smaller, more angularly shaped grains "floating" in microscopically fine material.

Clay and its compacted relation, shale, have an entirely different association of grain size and porosity. Clay, so fine it has no granular structure, is formed from platelets of minerals that are themselves the product of chemical (rather than mechanical) erosion. These settle to the seafloor in stacks and microscopic layers. They have high porosity but essentially no *permeability*, that is, no connection between pore spaces.

Mud is a term that embraces combinations of silt and clay. Intermixing of clay with silt severely limits the sediment's permeability.

Chemical rocks such as limestone and dolomite can have little *primary porosity*, that is, porosity when they are formed. So much limey mud and shell fragments have mixed around the corals that the rock formed without any original pore space. Occasionally, compaction and simultaneous circulation of seawater leaches calcium carbonate from the sediment, leaving pore spaces in the rock.

Often, some of the shell fragments themselves are dissolved. Further dissolution can occur by groundwater dissolving masses of the rock itself. These pore spaces in limestone and dolomite are called by the unpleasant name of *vugs*. They range in size from microscopic to cavernous. The Carlsbad Caverns, in New Mexico, are vugs, although their promoters would not likely advertise them that way. The vugs of the Yates Field, in West Texas, are an exceptional example of a limestone that was subjected to groundwater percolation and rock dissolution. As the discovery well was being drilled in Yates, the drill bit dropped more than 10 feet into a cavern filled with oil, reigniting vague memories of the fissure theory of the 1860s.

Typical oil and gas fields are contained in sedimentary rock formations that have porosities ranging from 5% to 30%. Igneous and metamorphic rocks have no porosity of importance to hydrocarbon accumulation except in the rare cases in which they have been severely fractured or weathered. More commonly, what fractures exist are not connected sufficiently for any worthwhile hydrocarbon accumulations to form. Sometimes metal ores fill the fractures, but more often, they have been filled with silica or calcium carbonate.

Connecting the spaces

It is in pore spaces that oil and gas may collect, but it is the connection of one pore space to another that permits the hydrocarbons to flow out of the rock and up the wellbore (fig. 2–14). *Permeability* is a measure of the degree of connectivity of the pore spaces of a rock.

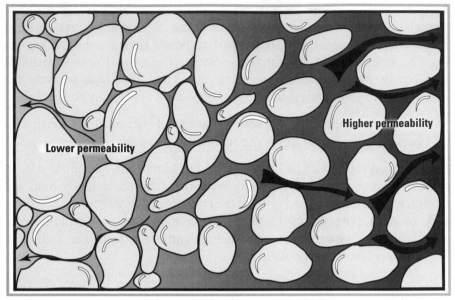

Fig. 2–14. High and low permeability. Where the rock is well sorted the flow paths are through large spaces and the permeability is high. The opposite exists where the grains are poorly sorted and the fluid must pass through small spaces along tortuous paths.

A French connection. Oil and gas exist in rock far more consolidated and cemented than the materials used by Henri Darcy, the French water department engineer, in the 19th century, when he measured water flow through a unit volume of sand. The resulting measure became known as a "darcy." In the petroleum industry, the common measure for the connectivity of pore space is one thousandth of a darcy, or a *millidarcy* (md). Commercial petroleum production typically comes from rocks having permeabilities ranging from a few darcies down to about 10 md —and in the least prolific cases, to just a few md.

The combinations of porosities and permeabilities are endless. Clay and shale, on the one hand, have very high porosity but no measurable permeability. Sands and sandstones, on the other hand, have permeabilities influenced by the combination of the degree of sorting of the grains and by the amount of cementlike material deposited, consolidating the grains into rock. A well-sorted sand has a preponderance of grains of the same size. A poorly sorted sand has a mixture of grain sizes. The more poorly sorted the sand is, the lower are both the porosity and the permeability.

Distortion

The plates that make up the crust of the Earth continuously move, albeit very slowly. At the margins or edges of the plates, the force of one plate against another can cause bending, rupturing, or deformation, building wonderfully scenic surface contours—mountains, valleys, escarpments, and trenches. Then there are the more subtle subsurface contortions of much interest to geologists—folds, faults, and unconformities.

Folds. The strong push-and-pull forces of plate tectonics created mountain ranges in some areas. In others, the forces were less pronounced, and the layers were less severely distorted into bends and folds. In many of these locales, hydrocarbons accumulated, patiently awaiting discovery.

Folding or bending of rock layers results in many shapes that have been named by geologists (fig. 2–15). Of great and historically early interest to the petroleum explorer is the *anticline* and its more symmetrical cousin, the *dome*, common structures in which hydrocarbons can be trapped as they migrate from their source toward the surface of the Earth.

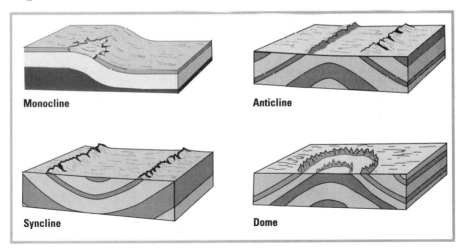

Fig. 2–15. Forms of folded strata. Rock layers deformed into a variety of shapes.

Faults. In the bending process, forces may cause the rock layers to break and the parts to become dislocated, a phenomenon named *faulting*. The movement of the ruptured rock layers can be up, down, or sideways (fig. 2–16), depending on the direction of the dynamic forces acting on the system. Often, the results are different types of rock layers on either side of the fault.

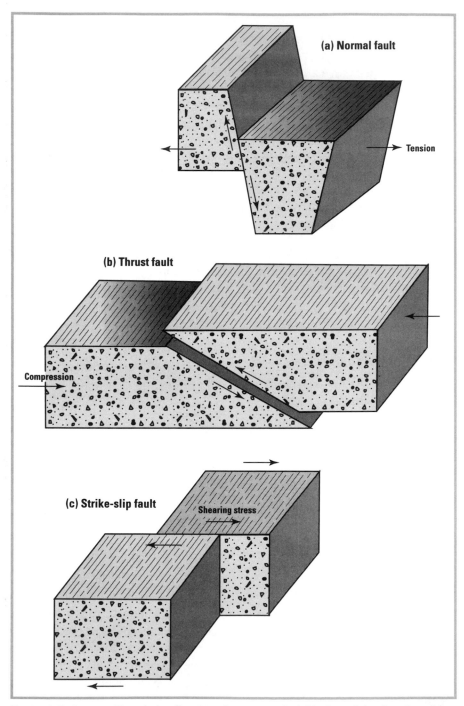

Fig. 2–16. Fault types. The relative direction of movement fault blocks and the direction of the deforming stresses.

Unconformities. A gap in the sedimentary sequence is an *unconformity* (fig. 2–17). When there is no sedimentation, there is erosion of the sedimentary surface as it stands above sea level. When that surface is covered once again by the ocean and the sedimentary process repeats itself, new sediments are deposited on the old surface. If the old surface is essentially flat, then geologists call it a *disconformity*. However, they are far more excited if the original surface has been tilted, deformed, and eroded before the new sedimentary sequences are deposited. This is called an *angular unconformity*, which phenomenon has resulted in some of the largest hydrocarbon traps because it may cover a relatively large area.

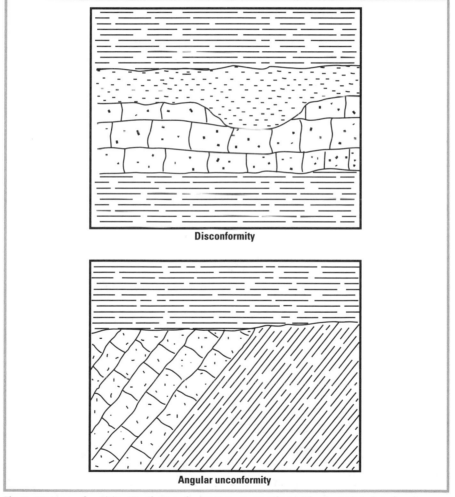

Disconformity

Angular unconformity

Fig. 2–17. Unconformities. On the top, layers top one another with an erosional surface in between. On the bottom, the layers have been tilted, eroded, and then overlain by horizontal deposits.

Traps. Hydrocarbon molecules are lighter than water. They move from where they formed, commonly called the *source rock*, through the water-filled pore spaces in rock, but always upward toward the surface. Unless they are intercepted, or *trapped*, the hydrocarbons spill out on the surface as tar pits, oil seeps, or natural gas seeps.

Folding, faulting, and unconformities are the principal mechanisms that trap and contain hydrocarbons, making commercial oil and gas production possible (fig. 2–18). Add to the list the distortions caused by salt domes, huge plugs of salt that intrude themselves up through overlaying sedimentary layers and the catalog of most common trapping mechanisms is about complete.

To intercept the migrating hydrocarbons and then contain them, a trap must be three-dimensional (3D). Rather than a simple fold or fault, there must be lateral closure—the trap has to stop the migration both vertically and horizontally. The anticline in figure 2–18 has to be folded along axes that are perpendicular, a structure shaped like an inverted canoe, to trap hydrocarbons. Similarly, a dome, a feature shaped more like an inverted cup, will also do the job.

Most adjacent rock layers vary in composition, for example, from sand to shale. Whatever the change, it is reflected in rocks with different porosities

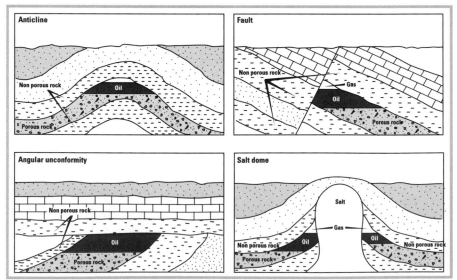

Fig. 2–18. Types of traps. Oil and gas migrate upward through porous rock and are finally trapped by non porous rock. The geometric relationship of the porous and non porous layers takes many forms; the most common are shown here.

and permeabilities in contact with each other. That in itself can produce the lateral, as well as vertical, trapping mechanism necessary to contain the migrating hydrocarbons, even without any fold or fault. Geologists call a sedimentary change that contains hydrocarbons a *stratigraphic trap or strat trap* (fig. 2–19).

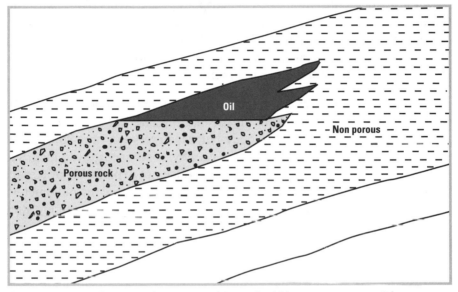

Fig. 2–19. Stratigraphic trap. Often a porous layer ends within non porous material, *e.g.,* sandstone in shale. The porous material is said to "pinch out."

Discovery

Origin ❯ Transformation ❯ Characterization ❯ DISCOVERY

The earliest oilmen understood little about lateral changes in sedimentary sequences, unconformities, or faults and folds below the surface. Even if they did, they could not identify or locate them. They simply drilled their wells where oil and gas seeps were prevalent. Before long, the more adventurous oilmen moved up the adjacent hills and drilled wells that were even more successful. The association of hills—and then of hills formed from anticlines—with hydrocarbon accumulations became the foundation of early exploration drilling based on geology. With this knowledge, wildcatters looked for anticlines and domes by mapping elevation changes in the rock strata visible on the surface as a way of locating their exploration wells.

Finding the trap

To find subtle surface indications of buried anticlines and other hydrocarbon traps, geologists began mapping the surfaces of the rock layers as they appeared to the eye. They used common surveyor tools and later, starting in the 1920s, aerial photography to expand their geologic overview. They concentrated on outcropping rock layers. Even with this limited surface information, they produced detailed maps depicting the topography of one or more rock layers. These maps gave them the temerity to postulate what lay beneath the Earth's surface and even to recommend a site for a wildcat well to test for an oil trap.

Mapping. Geologists continued to develop techniques to map the subsurface in a way analogous to the noninvasive testing that medical specialists examine the human body before surgery—x-rays, sonograms, MRIs, and CAT scans. Geologists and geophysicists used gravity meters to detect subtle discontinuities beneath the surface not explainable by surface mapping, especially around salt accumulations. They used magnetometers, devices that read the magnetic properties of the rock layers, as indicators of faulting and folding.

The Father of Surface Contours

In 1772, The British Royal Society underwrote an effort to weigh the planet Earth, using Isaac Newton's notion that a plumb bob hung near a mountain would incline slightly toward the mountain, affected by both the Earth's and the mountain's gravitational pull. One Nevil Maskelyene began surveying Schiehallion, a mountain in the Scottish Highlands chosen by the surveyor, Charles Mason (of Mason-Dixon Line fame.) After Maskelyne exhausted himself sizing up the mountain, he turned books with endless survey points over to a mathematician, Charles Hutton, to proceed with the calculation.

Pondering the confusing mass of numbers, Hutton noticed that if he used a pencil to connect the points on a map of equal altitude, it all became much more orderly. In a moment of simple brilliance, he had invented contour lines, giving him a visual sense of the shape of the mountain. This was the most modest of a series of geological insights that eventually earned Hutton his immortality as the Father of Geology. Oh, and in 1774, he finished his calculation of the mass of the Earth: 5×10^{21} tons, a figure within 20 percent of today's reckoning.

But like their medical counterparts, when geologists wanted to know more about the internal configuration, they had to get in situ data. Surgeons turn to the scalpel; geologists choose the drill bit. Not long after the industry began, drillers began to record what came out of their wells as they drilled and to measure the depths at which they found it. By World War II, thousands of wells had been drilled across the United States and Canada. Geologists tediously compared and correlated the data collected from neighboring wells.

Connecting the dots. Every well drilled encounters rock layers characterized by color, lithology, fossil content, hydrocarbon content, and more (the source of this data is discussed in chapter 6). Geologists compare and match the rock characteristics, layer by layer, of nearby wells or even wells several miles apart. They use the measured vertical depth to the top and bottom of each layer and note its relationship to sea level. With two wells (control points), there are just two different elevations. With at least three wells, they can make rudimentary 3D maps of a particular subsurface horizon (the top or bottom of a particular layer). The more wells compared (or as geologists say, the more control), the better and less controversial the maps become.

The map of the surface of a single rock layer, as viewed from above, shows the hills (anticlines), valleys (synclines), fault disruptions, and other discontinuities, but only of that surface. A map made at right angles to the surface map is called a *cross section* or *x-section*. It contains a sideways view of a particular stack or sequence of layered surfaces. This two-dimensional (2D) view of the Earth from a horizontal perspective helps visualize how the sedimentary layers are stacked and lie between two or more wells.

Rock layers are often so complexly arranged by folds, faults, and stratigraphic changes (lateral changes in lithology) that x-sections are invariably the product of the geologist's individual interpretation. (Some prefer the words *imagination* or *musing*, especially investors who have financed an unsuccessful well, or *dry hole*.) Geologists often get plenty of reinterpretation help from their colleagues, supervisors, and the investors evaluating the risk and potential of a drilling prospect.

By the late 1950s, most of the Earth's land surface had been examined and mapped. The obvious features, the trap indicators on the surface, had been drilled. At that point, a great unknown was what features lay hidden well below the ground's surface and in the vast area below the sea.

Enter seismic

The quantum leap in subsurface mapping started in the 1920s with the introduction of *seismic technology*. A second burst of capability followed with the introduction of digital computers in the 1950s.

Seismic involves four steps: *acquisition, processing, display,* and *interpretation*.

Acquisition. Onshore or offshore, the object of this phase is to collect seismic data that present a picture of the subsurface rock layers and their structuring. Onshore, a team of "jug handlers" place a series of sensors, called *geophones* or jugs, for thousands of yards around a spot, the *shot point*, at which they will release large shock waves. Offshore, seismic vessels like the one in figure 2–20 tow *streamers*, plastic tubes that extend up to 30,000 feet in length, behind the boat. The plastic tubes contain *hydrophones*, thousands in each streamer. The vessel also tows air guns, the seismic source, a short distance behind the boat. These guns are filled with compressed air, and when they open abruptly, they release the air with a bang, like a popped balloon. As that happens, the classic seismic data gathering takes place (fig. 2–21).

Onshore, in populated areas, the seismic source is a truck-mounted plate that is pounded on the ground in a series of vibrations lasting up to 20 seconds. Where civilization has yet to arrive (foreign deserts and swamps) exploded dynamite provides the energy source.

Fig. 2–20. Seismic vessel. The wake far behind the vessel is formed by the hydrophone streamers being towed. Courtesy of Veritas DGC.

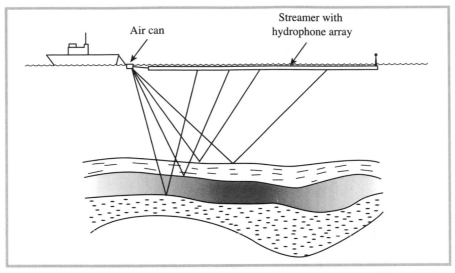

Fig. 2–21. Offshore seismic data acquisition. Energy waves from the air can travel downward and bounce back from numerous levels to the hydrophones, the receiving devices.

As the sound wave travels down through the water or directly into the ground, it reflects or echoes off the boundaries between various layers of rock, back to the surface, where the hydrophones or geophones (the 'phones) pick up the echo. Different 'phones record the bounced seismic signal from different angles. The deeper the layer—and this is the essence—the longer the echo takes to reach a 'phone. Moreover—and this is critical as well—the various layers of rock in the subsurface have different acoustical properties, depending on their lithology and *fluid content* (what is contained in their respective pore space). The unique composition, density, and contents of each rock layer affect velocity of the echo differently, thus enabling seismic interpretation.

Individual *seismic records* appear on displays as black wiggly lines. On the record in figure 2–22, the intensity (or amplitude) of each line (geophysicists use the term seismic trace) measures the strength of the reflected signal.

The number of recorded sounds can add up to mind-numbing and incomprehensible amounts. Some surveys capture and store several terabytes of data. (Measuring information is described in increasing numbers as bytes, kilobytes, megabytes, gigabytes, and then terabytes, or 10^{12} pieces of information.)

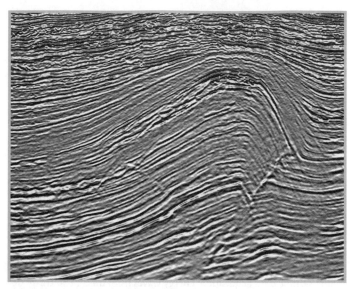

Fig. 2–22. 2D seismic display of an anticlinal feature (refer to color plate for hydrocarbon indicators shown in red). Courtesy Landmark Graphics.

The seismic travel time for a reflection at a depth of 10,000 feet varies from about 1.5 to 3 seconds. Seismic records typically have 12–20 seconds worth of data. The reflected sound waves are converted into a digital format by using special modules in the 'phone cables, and those data are carried by cables to a recording room on the seismic boat or recording truck.

If the 'phones are strung out in a straight line, the seismic record comes in 2D format. The reflections come from below, but not from the side. To create 3D records, seismic vessels or jug handlers arrange parallel strings, as the vessel in figure 2–20 has done. During processing, the parallel records are transformed into a continuous 3D block of data, often called a *seismic cube*.

Processing. Some processing takes place at the recording site to organize the data, but most of it, plus all subsequent reprocessing, is performed using mainframe computers in the processing centers of either the service company or the operator.

Data manipulation includes elimination of bad records, corrections for unwanted shallow surface effects, and reduction of the effect of multiple reflections (wave fronts that bounce around within geologic layers and then back to the surface). This work is done using a series of software programs developed from mathematical theory about seismic wave fronts traveling through the subsurface. Ultimately, the data is assembled in eye-friendly formats in the form of wiggle traces that are lined up and represent reflections from rock layers.

Display. Seismic work leans heavily on visualization. Displays of 2D vertical slices like the one shown in figure 2–22, either as hard copy or on computer screens, provide first looks at the geology. Horizontal slices of the data cube, usually called *time slices*, can also be displayed.

Early on, interpreters found that the amount of data in black-and-white seismic lines overwhelmed the human eye. However, the eye can handle wide-ranging variations in color. Interpreters, even as late as the 1970s, used colored pencils to add another dimension to their displays. That sometimes led to disparaging but good-natured remarks about the sophistication of seismic interpreters. Eventually, software programs differentiated amplitudes, spacing, and other characteristics of the seismic reflections with color arrays chosen at the pleasure of the interpreter (fig. 2–23).

Now, 3D seismic data, along with geologic and engineering data, are displayed in visualization rooms, like that in figure 2–24. Images of the subsurface are shown on large high-fidelity screens and appear on one or more walls to create the illusion for the viewer of being present in the

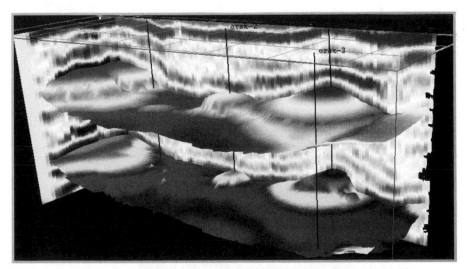

Fig. 2–23. 3D display of an oil and gas field (see color plate). Courtesy Landmark Graphics.

subsurface looking at the data from different angles, sometimes using special 3D glasses. A 3D cube like that in figure 2–25 can be rotated on a computer screen to get a different view of the seismic image of the subsurface.

Interpretation. All this preparation is just a prelude to the final step: interpretation and making economic decisions that may lead to drilling. Teams that can include geophysicists, geologists, petrophysicists, and other professionals bring their special knowledge to the interpretation, as they

Fig. 2–24. Seismic display in a visualization room (see color plate). Courtesy Landmark Graphics.

Fig. 2–25. 3D seismic block (see color plate). Courtesy Landmark Graphics.

Color Plates

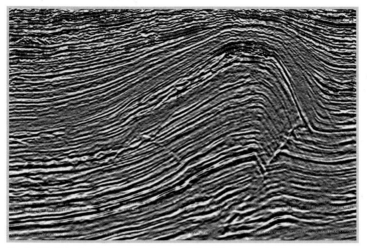

Fig. 2–22. 2D seismic display of an anticlinal feature, with hydrocarbon indicators in red. Courtesy Landmark Graphics.

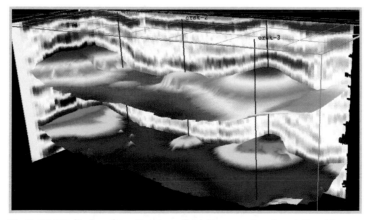

Fig. 2–23. 3D display of an oil and gas field. Courtesy Landmark Graphics.

Fig. 2–24. Seismic display in a visualization room. Courtesy Landmark Graphics.

Fig. 2–25. 3D seismic block. Courtesy Landmark Graphics.

search for the reservoir, the trap, and direct indicators of hydrocarbon presence. That calls for close interaction to relate seismic data to the geologic and geophysical knowledge of the area.

But the operative word remains—interpretation. Seismic mapping still remains in good part art. Geophysicists have developed mathematical algorithms to account for numerous distortions of the waves as they bounce around unconformities, folds, faults, different lithologies, and other subsurface vagaries. Even a slight error in the algorithm can cause what seemed to be a promising structural closure—the trap, one of the essentials for success—to vanish when tested by the drill bit. Still, seismic work plays a critical role in the decision to drill the wildcat well and even delineation or development wells.

And On...

A suitable container is only one of the essential parts of a commercial oil and gas accumulation. The next logical question is, "How do the hydrocarbons actually get there?"

3

What's in the Container?
The Prize

For four-fifths of our history, our planet was populated by pond scum.

—J. W. Schopf (1942–), Astrobiologist,
University of California, Los Angeles

Of course, people in the oil patch don't call it a container—they say "reservoir." But whatever name they use, it contains the fluids they are interested in, oil and natural gas, and some they are not, water and other miscellaneous gases.

A better understanding of the nature of the reservoir's contents comes from running through some of the factors that have varied over the history of the container—pressure, temperature, and time. These are crucial to the formation of hydrocarbons. They have an impact both on the kind of hydrocarbons that are present and on the way operators can produce them.

Determinants

Pressure

The load placed on an object's surface area at any point below the Earth's surface is measured in *pounds per square inch* (psi). As *overburden* builds up by sediment deposition, or as the Earth's crust deforms, the additional weight increases the pressure. Sediment or rock generally exerts a pressure of about 1 psi per foot of depth, or 5,000 psi at

5,000 feet. This force, *rock pressure*, sometimes called *overburden pressure*, is supported by the rock itself and is transmitted from grain to grain (sand or silt grain to sand or silt grain).

The pore spaces between the supporting grains have their own, independent pressure-causing mechanism. The pressure in the pores right next to those grains, known as *hydrostatic pressure*, is equal to the weight of the column of saltwater that fills the spaces above the reference point. It is also measured in psi. Hydrostatic pressure generally increases at a rate of 45 psi for every 100 feet. At a depth of 5,000 feet, the hydrostatic pressure would be 2,250 psi.

The overburden pressure can sometimes become so great it deforms or even crushes the supporting grains. Normally, the fluid is squeezed out. Sometimes the fluid is contained by some impermeable seal like shale and cannot escape. That leads to increasingly high or *abnormal pressures*. These abnormal pressures always concern drillers, who worry about the prospect of a blowout when the drill bit penetrates an overpressured sequence of sediments. That warrants more discussion in chapter 5 about containment and protection when abnormal pressures are encountered during drilling.

Temperature

Chapter 2 touched on the high temperatures of the Earth's core. The Earth's surface, in contrast, has only its atmosphere to shelter it. The sun's rays keep it at habitable and, in comparison, moderate temperatures. Between the two points, from the surface to the core, the subsurface gets increasingly warmer. Worldwide, *the geothermal gradient* (the rate at which the temperature increases with depth), averages about 15°F per 1,000 feet of depth below the surface. The increase has to be added to the average annual temperature at the surface. Calculating the temperature of a reservoir at 5,000 feet in the Powder River Basin of Wyoming, for example, would be the 65°F average surface temperature plus 5 × 15°F, or 140°F. Offshore, at depths of around 5,000 feet, the water temperature is about 33°F, so the temperature at the bottom of a well 25,000 feet below sea level would be about 333°F.

Variations abound. For example, near hot spots like Yellowstone Park, the subsurface temperature is warmer. Where the Earth's crust is being pushed beneath a continental plate and the temperature has not yet adjusted, it is cooler.

Time

Accumulations of animal and vegetable material (*biogenic mass*) can be converted to hydrocarbons in a laboratory by applying an external source of heat. In nature, the heat necessary to effect the conversion comes after subsidence. As biogenic material is buried beneath the Earth's surface by sedimentation or crustal movement caused by plate tectonics, temperatures rise, just because the material is getting closer to the Earth's center. Without the benefit of laboratory conditions, the natural process takes a bit longer—several million years or so. But the Earth is 4.5 billion years old, and earth scientists think primitive forms of life started forming a billion years ago. By 500 million years ago, the seas contained a sufficient accumulation of living animal and vegetable organisms to become the source of hydrocarbon generation, awaiting "only" the geologic processes to bury the sediments from these ancient seas and to cook them a while.

The Contents

For the first 10–50 feet below the Earth's surface, the ground is relatively dry. It becomes increasingly moist until at some point, called the water table, it becomes 100% water saturated. Water can move freely to and fro in this region. The pore spaces, those intergrain voids in all rocks, contain fluids below the water table. In almost all cases, the fluid is water, freshwater down to about 1,100 feet, mostly salty water below that. Hydrocarbons exist where there is a trap to prevent their floating away—migrating to the surface. But more on that later.

Saltwater

As sediments accumulated, they trapped the saltwater in the sediment's pores as formation water. Normal seawater contains 25,000–35,000 parts per million (ppm) of sodium chloride. As the sediments accumulated into layers and their weight compressed the pore spaces below, the saltwater was pushed out—just like squeezing a wet sponge, albeit somewhat slower. As the water moved from one porous layer to another, it sometimes picked up additional salt by dissolving salt deposits that could have been left behind by evaporation of ancient seas (evaporite deposits). That explains why the salt content of formation water varies from 50,000 to more than 110,000 ppm. At the high end, the formation water, called brine at that elevated concentration, is saturated with salt—no more can be dissolved in it.

Hydrocarbons: oil and gas

Understanding the origins of hydrocarbons begs a discussion of what they are. Hydrocarbons are combinations of carbon and hydrogen atoms connected into molecules of countless combinations and configurations. Miscellaneous atoms, sulfur, nitrogen, metals, and others sneak into some hydrocarbon molecules in minor amounts. Depending on how much, they can adversely affect both the production and processing of oil or gas.

The Chemistry

The branch of science called organic chemistry deals with hydrocarbons, the combinations of carbon and hydrogen atoms, and presents a sometimes insurmountable hurdle to college students aspiring to the medical profession and engineering careers. By default, many end up as lawyers and business persons.

Petroleum engineers have only a moderate concern with organic chemistry. Still, they and other people interested in production ought to know something about the chemistry of hydrocarbons.

The simplest hydrocarbon, *methane*, has four hydrogen atoms and one carbon atom connected to each other (fig. 3–1). The key to that and other, more complicated molecules is the *valence* of the two types of atoms. A simple way to think of valence is the number of other atoms that a particular type of atom wants to connect to. Hydrogen has a valence of one; carbon has a valence of four. For that reason, the carbon atom in methane connects to four hydrogen atoms, and each hydrogen atom connects to one carbon atom. The connections are called *bonds*. Almost all the molecules that make up oil and gas are built on this principle.

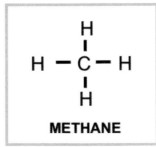

Fig. 3–1. Methane molecule

The next-simplest molecules are ethane, propane, and butane (fig. 3–2), in which the valence requirements are all satisfied. As these molecules get successively bigger, they follow the formula $C_nH_{2n} + 2$, and are called *paraffins*. The longer paraffins, in which there are more than 20 carbon atoms, are waxy solids at room temperature. Crude oils that have a lot of paraffins can cause *waxing* problems when the paraffins precipitate as solids and stick to the walls of pipes, valves, and vessels.

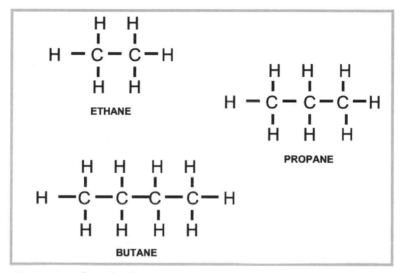

Fig. 3–2. Paraffin molecules

But not all molecules stretch out in straight lines. *Naphthenes* wrap themselves around into cyclic shapes like that of the cyclohexane molecule shown in figure 3–3. Their general formula is C_nH_{2n}, but all the carbon and hydrogen valences are still satisfied.

Combinations—untold numbers of them—can arise from connecting naphthenes and paraffins, such as the dimethyl cyclohexane in figure 3–4, and that makes studying the chemistry of oil and gas endless.

Two other general forms of molecules complete the oil and gas story but undermine the principles. Molecules called *aromatics* (fig. 3–5) are missing some hydrogen atoms that should satisfy the carbon valence of four. Instead, the carbon atoms have *double bonds* in the ring, as if two bonds take the place of one, satisfying the carbon atom's need to have four connections. Like naphthenes, aromatics can team up with paraffins to create countless combinations.

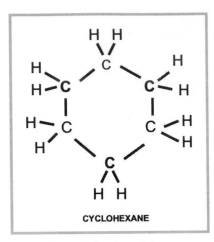

Fig. 3–3. Naphthene molecule: cyclohexane

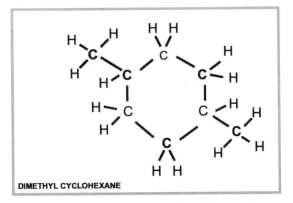

Fig. 3–4. A complex combination of naphthene and paraffin groups: dimethyl cyclohexane

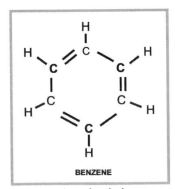

Fig. 3–5. An aromatic molecule: benzene

Finally, *asphaltenes*, complicated molecules with structures that look like balled-up chicken wire, have an abundance of aromatic shapes connected to each other. Asphaltenes have numerous carbon and relatively few hydrogen atoms and formulas like $C_{57}H_{32}$.

Crude oil is made up of 100,000–1,000,000 different types within these four categories of molecules, some more common than others. The American Petroleum Institute, the U.S. Bureau of Mines, and other august groups classify crude oils using this chemistry-based characterization:

- paraffinic crudes
- naphthenic crudes
- aromatic crudes—including asphaltic crudes, as a subcategory

The casual observer might conclude that paraffinic crudes are made up of paraffins, but the truth is that paraffinic crudes are predominantly paraffinic—they still contain plenty of naphthenes and aromatics, as much as 50%. The same story is true for the other categories. Still the paraffinic-naphthenic-aromatic classification helps refiners know what kinds of processing they will have to do and what types of products they can easily make.

Composition

While as many as a million different types of molecules make up a crude oil, only a few make up natural gas. Figure 3–6 shows six different classifications for oil and gas. These *commercial* names are used for the streams *as they are produced* (chapter 7 will deal with the importantly different nomenclature used by petroleum engineers to categorize hydrocarbons *in the reservoir*):

- *Dry gas* is a natural gas stream that consists almost entirely of methane, with possibly a few percent ethane and propane.

- *Wet gas* is a natural gas stream that consists of a high percentage of methane (80–90%), plus *natural gas liquids,* ethane, propane, butane, and natural gasoline, and small amounts of other constituents, as shown in table 3–1.

- *Condensate* is a very light crude oil–type hydrocarbon that comes from a well producing predominantly natural gas (a gas well). Condensate generally has some natural gas liquids in it.

- *Light crude oil* probably has a few percent natural gas liquids.

- *Heavy crude* has little or no natural gas liquids and a high percentage of the heavy hydrocarbons. Sometimes the heaviest of the heavy crudes have to be heated to make them fluid enough to pump out of the ground and through a pipeline.

- *Bitumen* is a composition of heavy hydrocarbons, including the very complex asphaltenes. Together, they form solids at ambient temperatures. Bitumen is a hard substance to handle. In the reservoir, bitumen is a thick, sluggish fluid like road asphalt and has to be extracted from the ground with much heat and effort and hauled away to a facility that can upgrade it to a useful product.

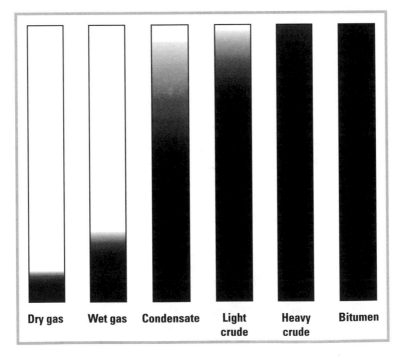

Fig. 3–6. Produced hydrocarbon characteristics. Dry gas is made up almost entirely by the lightest hydrocarbon: methane. Bitumen is comprised of large, complex molecules. Condensate differs from light crude usually by a lower percent of heavier molecules.

Table 3–1 Typical Constituents of Natural Gas

		Percent
Methane	CH_4	70-98
Ethane	C_2H_6	
Propane	C_3H_8	0-20
Butane	C_4H_{10}	
Carbon dioxide	CO_2	0-8
Oxygen	O_2	0-0.2
Nitrogen	N	0-5
Hydrogen sulfide	H_2S	0-5
Rare gases	A, He, Ne, Xe	trace

The constituents of dry gas and wet gas stay in the gaseous state at ambient temperature and have to be kept under pressure to be contained. The U.S. industry measures natural gas stream volumes in standard cubic feet. The energy content of a standard cubic foot of natural gas can vary, depending on the amount of components besides methane. A measure of the energy content of natural gas is *BTU per standard cubic foot* (natural gas typically has 1000 BTUs per standard cubic foot). A British thermal unit (BTU) is the amount of heat required to increase the temperature of a pound of water by 1°F. One other standard measure: many companies report their oil and gas statistics in *barrels of oil equivalent* (BOE). The conversion of natural gas to BOE is typically 5.8 million BTUs of gas per BOE.

Natural gas had a slow commercial start in most places. When substantial amounts of gas accompanied oil production, "black oil" producers considered it a nuisance and condescendingly called it *associated gas* (and in the old days often proceeded to flare it). But no one calls the liquids coming from a gas well "associated oil." They are condensate or natural gas liquids, depending on whether they are separated at the well site (condensate) or at a gas plant (natural gas liquids) that could be some distance away.

Properties

It is impractical to characterize hydrocarbons by their actual exact components, so a number of properties have come into use to describe how the hydrocarbon mixtures behave.

Gravity—or more exactly, *specific gravity*—is a measure of the weight of a given material compared to the weight of an equal volume of water at a standard pressure and temperature. For oil, the simpler and smaller the molecules contained in the mixture are, the lighter is the oil. Conversely, the more complex and larger the molecules are, the heavier is the crude. The oil industry's standard scale for measuring gravity, adopted by the American Petroleum Institute in the early 20th century, is *degrees API*, or °API. The formula relating °API to specific gravity is

$$°API = (141.5 \div \text{specific gravity}) - 131.5$$

The logic of this arcane formula has been lost in antiquity. Nevertheless, all industry uses this measure to specify the various types of crude oil. Table 3–2 shows one such categorization

In industry, the usage of the °API categories in table 3–2 is lax. Clearly, some proper names, the ones containing the words "light" and "intermediate," do not relate to the classification scheme and can mislead the unwary buyer.

In many areas, the °API number of a specific crude oil is used as an index to adjust the prices of that crude oil from others in the same area but with slightly different characteristics. Lighter crudes almost always command a higher price than heavier crudes.

Viscosity measures the fluid's resistance to flow, its "thickness." Molasses and tar are viscous and have high viscosity numbers. Water has a low number. Normal light to medium crude oils pour easily, like water, and more important to the petroleum engineer, move through the pore spaces more easily. Viscosity is measured in centipoises (cp), but production engineers think of it in the uninspiring terms of "low vis" and "high vis."

Pour point, color, and odor are other useful ways to characterize a crude oil. Pour point measures the lowest temperature at which a crude oil will flow, just before it starts to turn into a solid. Generally, the higher the paraffin

Table 3–2. Crude oil classifications

°API		Examples
<10	Bitumen	Canada's Athabasca (8 °API) and Venezuela's Orinoco (8–10 °API)
10-20	Heavy Crude oil	California San Joaquin Valley Heavy (13.4 °API) and Venezuela's Bachaquero (17.2 °API)
20-35	Medium crude oil	Arab Light (34 °API), Venezuela's Tia Juana Light (31 °API) or West Texas Sour (34 °API)
35-50	Light crude oil	West Texas Intermediate (40 °API), Brent (38.5 °API) and Nigeria's Bonney Light (37 °API)
50+	Condensate	Indonesia's Arun (53.9 °API), Bossier Parish (Louisiana) condensate (58 °API), and "Okie" gas in the 1930s because of its ability to be burned as gasoline in a car

content of a crude is, the higher will be the pour point. (Lower pour point is better than higher pour point, of course, because pumpers want a crude to continue to flow when the snow starts to fall.)

Crude oils vary from nearly colorless (very light crudes) to greenish-yellow to reddish to black (generally the heavy crudes). Various crude oils smell like gasoline (sweet crudes) or rotten eggs (sour crudes) or have a sickly fruity smell (aromatic crudes).

Impurities

All sorts of cats and dogs can turn up in a hydrocarbon reservoir, whether it contains oil or gas or both. Along with natural gas and its natural gas liquid constituents, other gases can present themselves (table 3–1). Carbon dioxide (CO_2), oxygen (O_2), nitrogen (N_2), and hydrogen sulfide (H_2S) are the most common. Not only is hydrogen sulfide a lethal gas in very small concentrations, but burning it creates a pollutant. That gives two good reasons why environmental laws and safety considerations require it be removed. Excess carbon dioxide has to be eliminated because it can cause corrosion of transportation and processing equipment.

A gas stream can contain minor amounts of the rare gases—argon, helium, neon, and xenon. Sometimes they are in sufficient quantities to make separation and recovery commercially attractive.

Crude oils can have the same contaminant problems. Most serious for crude oil is the presence of sulfur and metals. Sulfur can reside in crude oil in the form of dissolved hydrogen sulfide, or it can be sulfur atoms chemically attached to the hydrocarbon molecules like thiophene (C_5H_5S). Depending on the country, environmental regulations require almost all the sulfur be removed from refined products before they can be sold. Special units in refineries are required in order to remove it, and as a result, the more sulfur there is in the crude, the lower will be its value to refiners.

Metals, such as vanadium, nickel, and copper, can damage the catalysts used to process a crude oil in refineries and can also cause debits to the values of crude oils so contaminated.

Fingerprints

By the time an oil accumulation has found its present-day resting place, it has followed a complex process of creation and migration. As a result, each oil has a unique "fingerprint," including the characteristics just described. When there is an oil spill offshore, chemical analysis can often point to the producing reservoir that is the culprit, just as DNA may be used to associate a felon with a crime scene.

The Origin of Hydrocarbons

Two contemporary theories deal with the origin of hydrocarbons, *biogenic* and *abiogenic*. The first, the theory supported by most geologists and other scientists in oil companies and used so far in this book, purports that organic matter, mostly microscopic animal life and some plant life, was the source; the second attributes the origins to primordial sources and the processes that formed the Earth.

Biogenic
The biogenists theorize that in some locations, sea-dwelling micro-organisms in great abundance fell along with clay particles to seafloors. These biogenic rich clays were covered with succeeding layers of sediment.

As the burial process took place, the clay intermixed with the biomaterial and became shale, the source rock for hydrocarbons. Not all shale contains enough biogenic material to qualify as source rock. There is room for debate about how much organic material is necessary for hydrocarbon generation, but some scientists suggest 3–8%.

With continued sedimentation or other geologic movement downward, the temperatures rose (recalling that the closer to center of the Earth, the higher the temperature), and the microorganisms began to "cook." Over time, heat decomposed most of the organic material, and a small proportion polymerized and then split into hydrocarbons of various sizes and shapes. After they formed, many migrated out of the source rock, either to traps where they awaited the drill bit or they escaped into the atmosphere.

The conversion to hydrocarbon began when the temperature rose to about 180°F. If and when the sediments were buried deep enough for the temperature to rise beyond 295°F, the process created smaller molecules, predominantly methane. If the temperature rose much above 450°F, everything decomposed to carbon dioxide and water, and no hydrocarbons remained.

From all this, geologists surmise a lot about the depths at which oil and gas formed. The conditions were ideal for oil formation at 7,000–15,000 feet and for gas at 15,000–25,000 feet. They call these ranges *oil windows* or *gas windows of formation*. An announcement by a company that it intends to drill to 25,000 feet tells a lot about its objectives.

Once the hydrocarbons have formed and migrated to a trap and accumulate, tectonic forces can continue to bend and warp the rock strata that formed the trap. These forces can have moved the hydrocarbon accumulations up and down, in and out of the hydrocarbon window. That is why oil and gas can be found outside the oil and gas windows. In fact, moving into the "too hot" zone, geologists surmise, may cause oil to convert to gas and gas to cook off from some reservoirs, leaving only a residue of the former hydrocarbons.

Abiogenic

The abiogenic proponents have more expansive views of hydrocarbon formation. In one version, they allege that hydrocarbon formation took place in the Earth's mantle, 60–180 miles below the surface (fig. 3–7). The mechanism still required organic chemistry, but conversion not of living matter but of other molecules that had carbon in them. Iron carbide and water percolating from the surface reacted to form acetylene which grew into longer or more complex hydrocarbons. Another abiogenic school purports that gaseous hydrocarbons, mostly methane, existed in the Earth's atmosphere, just as it does today around Saturn and Jupiter. These hydrocarbons rained down and formed the basis for today's hydrocarbon deposits—or at least some of them.

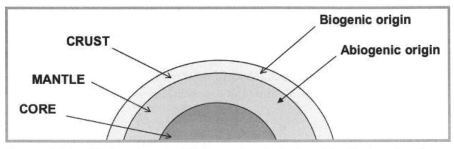

Fig. 3–7. Origins of hydrocarbons. Each if the two theories for the origin of hydrocarbons have a unique location within the Earth for their source.

Meteorites and deposits of congealed magma that contain traces of hydrocarbons found over the past half-century have given continuing momentum to the minority of scientists who support the abiogenic theory. Further, they have demonstrated in the laboratory the ease with which hydrocarbons can be created from inorganic (abiogenic) sources.

The abiogenic chemical transformation process that produces the variety of hydrocarbons now evident is about as complex as for the biogenic material, but it requires much longer migration to the reservoirs where hydrocarbons have already been found. If the abiogenicists are right, the amount of hydrocarbons yet to be found are beyond current comprehension, at places generally considered unconventional and un-suitable, mainly in targets considered too deep.

And On...

Regardless of who is right, the biogenic or the abiogenic devotees, oil and gas end up in reservoirs and have to be tapped with the drill bit. And that's the next subject.

4 Yours, Mine, or Theirs? Ownership

If you can walk with crowds and keep your virtue...
Yours is the Earth and everything that's in it.

—Rudyard Kipling (1865–1936), *If*

If a poll were taken among Americans, no doubt it would reflect that tragically few realize the difference between U.S. property rights and the rights of landowners in the rest of the world. They would be unaware, in particular, of Americans' inalienable rights to exploit the minerals beneath the land that they possess. And that includes hydrocarbons.

With few exceptions (Canada is one), governments outside the United States own the minerals in their countries, both onshore and offshore, without regard to surface ownership. In contrast, U.S. residents can own both the surface and the minerals underlying it, and they may sell or lease those rights to whomever they please and under whatever terms suit their fancies. Like other countries, offshore U.S. lands are owned by the government of the contiguous coastal state or by the federal government, depending on the distance from the shoreline.

Ownership

U.S. homeowners typically own a plot of land within a development or subdivision. The land either was built upon, may have been bought from a farmer or a rancher, sold to a developer, and resold several times before the final developer subdivided the land into individual plots and subsequently sold them, with or without houses, to individuals. During the transfer of the title to the land, likely as not, the mineral ownership was *severed* (separated) from the surface ownership. As a consequence, homeowners typically have little to say about the presence or absence of minerals beneath their homesteads. Someone else owns the mineral rights.

In a rural setting, the mineral ownership is usually quite different. The farmers, ranchers, or acreage owners customarily own the rights to the minerals beneath the surface. If they do not, it is because somewhere in the chain of ownership a surface owner retained (all or part of) the mineral rights when he or she sold the property. Often, ownership of rural lands has been complicated by inheritance. Many farms and ranches are owned in undivided interests by the heirs of the original owners. They own it collectively. In that case, every owner must sign any contract involving the disposition of the property.

Royalties

Ownership of the minerals does not necessarily go hand in hand with the knowledge of what lies below the surface or the financial resources to explore and develop a "find." In normal circumstances, a *landman* (a professional representing an individual, a group of investors, or a company) approaches the mineral owners with an offer to *lease* their rights. This, in essence, is an offer to acquire a major portion of the mineral rights while reserving a *royalty interest*, a percentage of the revenues from oil and gas production, for the landowner. In exchange, the acquiring company assumes an obligation to drill or otherwise explore the property. Typically, the agreement between the company and the mineral owners includes a time limit, often three years.

The acquiring parties are known as the *lessees* or working-interest owners, and the mineral owners are known as the *lessors* or royalty-interest owners. The lessees may continue to acquire other mineral leases until they have accumulated enough to drill the wildcat well and to follow up any discovery with development wells. It is the development wells that will return the profit from the venture.

When a landman knocks on the door of the mineral owner, the discussion will revolve around a number of points:

- signing bonus for the mineral owner

- royalty for the mineral owner, usually between one-eighth and one-quarter of the revenue from the venture, but sometimes as high as one-half

- obligations, an important one being the period in which seismic or drilling operations must commence, often three years, but perhaps only 60 days

- primary term, the period in which production must begin, normally 5 or 10 years

- amount of the mineral owner's acreage that will be in the prospective drilling unit, which determines the mineral owner's share of the revenues from the first well

- how much of the prospective mineral owner's land will eventually be "held" by production—the acreage assigned to the producing wells

- which party pays for treating costs to make the produced hydrocarbons marketable, which can be different for oil and gas

- when, and under what circumstances, the leased interest will *revert* (be returned) to the mineral owner

If the landman is sufficiently persuasive, the mineral owner signs a lease agreement. Figure 4–1 shows page one of a so-called standard lease (available on the Web site of the American Association of Professional Landmen). Of course, there is no such thing as a standard lease, and some of the negotiable items are highlighted. No explicit clause is even included in that form for a signing bonus.

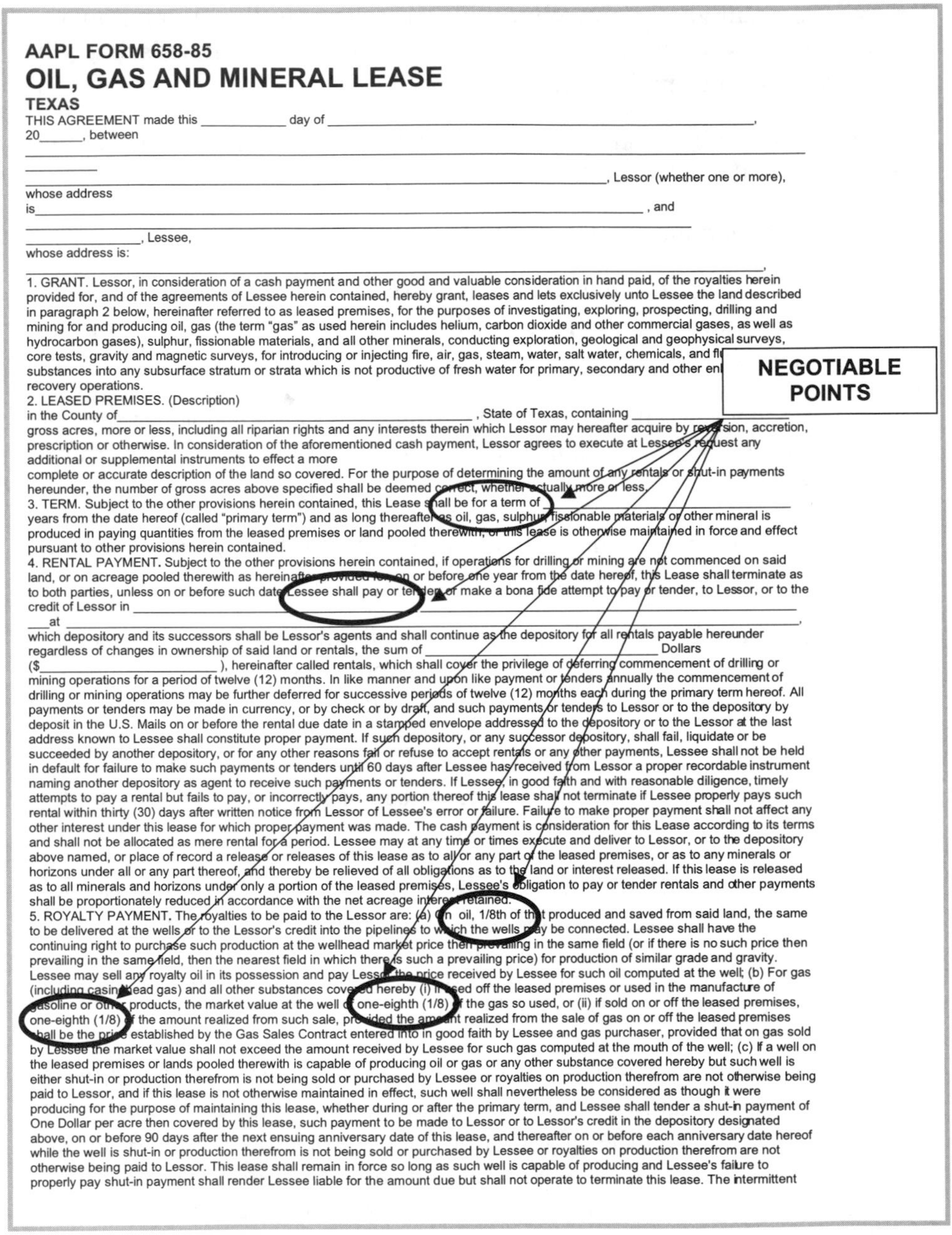

AAPL FORM 658-85

OIL, GAS AND MINERAL LEASE

TEXAS

THIS AGREEMENT made this _____ day of _____,
20_____, between

_____

_____ , Lessor (whether one or more),

whose address
is_____ , and

_____ , Lessee,

whose address is: _____

1. GRANT. Lessor, in consideration of a cash payment and other good and valuable consideration in hand paid, of the royalties herein provided for, and of the agreements of Lessee herein contained, hereby grant, leases and lets exclusively unto Lessee the land described in paragraph 2 below, hereinafter referred to as leased premises, for the purposes of investigating, exploring, prospecting, drilling and mining for and producing oil, gas (the term "gas" as used herein includes helium, carbon dioxide and other commercial gases, as well as hydrocarbon gases), sulphur, fissionable materials, and all other minerals, conducting exploration, geological and geophysical surveys, core tests, gravity and magnetic surveys, for introducing or injecting fire, air, gas, steam, water, salt water, chemicals, and fl__ substances into any subsurface stratum or strata which is not productive of fresh water for primary, secondary and other en___ recovery operations.

NEGOTIABLE POINTS

2. LEASED PREMISES. (Description)
in the County of_____ , State of Texas, containing _____
gross acres, more or less, including all riparian rights and any interests therein which Lessor may hereafter acquire by reversion, accretion, prescription or otherwise. In consideration of the aforementioned cash payment, Lessor agrees to execute at Lessee's request any additional or supplemental instruments to effect a more
complete or accurate description of the land so covered. For the purpose of determining the amount of any rentals or shut-in payments hereunder, the number of gross acres above specified shall be deemed correct, whether actually more or less.
3. TERM. Subject to the other provisions herein contained, this Lease shall be for a term of _____
years from the date hereof (called "primary term") and as long thereafter as oil, gas, sulphur fissionable materials or other mineral is produced in paying quantities from the leased premises or land pooled therewith, or this lease is otherwise maintained in force and effect pursuant to other provisions herein contained.
4. RENTAL PAYMENT. Subject to the other provisions herein contained, if operations for drilling or mining are not commenced on said land, or on acreage pooled therewith as hereinafter provided for, on or before one year from the date hereof, this Lease shall terminate as to both parties, unless on or before such date Lessee shall pay or tender or make a bona fide attempt to pay or tender, to Lessor, or to the credit of Lessor in _____ ,
___at _____
which depository and its successors shall be Lessor's agents and shall continue as the depository for all rentals payable hereunder regardless of changes in ownership of said land or rentals, the sum of _____ Dollars
($_____), hereinafter called rentals, which shall cover the privilege of deferring commencement of drilling or mining operations for a period of twelve (12) months. In like manner and upon like payment or tenders annually the commencement of drilling or mining operations may be further deferred for successive periods of twelve (12) months each during the primary term hereof. All payments or tenders may be made in currency, or by check or by draft, and such payments or tenders to Lessor or to the depository by deposit in the U.S. Mails on or before the rental due date in a stamped envelope addressed to the depository or to the Lessor at the last address known to Lessee shall constitute proper payment. If such depository, or any successor depository, shall fail, liquidate or be succeeded by another depository, or for any other reasons fail or refuse to accept rentals or pays, Lessee shall not be held in default for failure to make such payments or tenders until 60 days after Lessee has received from Lessor a proper recordable instrument naming another depository as agent to receive such payments or tenders. If Lessee, in good faith and with reasonable diligence, timely attempts to pay a rental but fails to pay, or incorrectly pays, any portion thereof this lease shall not terminate if Lessee properly pays such rental within thirty (30) days after written notice from Lessor of Lessee's error or failure. Failure to make proper payment shall not affect any other interest under this lease for which proper payment was made. The cash payment is consideration for this Lease according to its terms and shall not be allocated as mere rental for a period. Lessee may at any time or times execute and deliver to Lessor, or to the depository above named, or place of record a release or releases of this lease as to all or any part of the leased premises, or as to any minerals or horizons under all or any part thereof, and thereby be relieved of all obligations as to the land or interest released. If this lease is released as to all minerals and horizons under only a portion of the leased premises, Lessee's obligation to pay or tender rentals and other payments shall be proportionately reduced in accordance with the net acreage interest retained.
5. ROYALTY PAYMENT. The royalties to be paid to the Lessor are: (a) On oil, 1/8th of that produced and saved from said land, the same to be delivered at the wells or to the Lessor's credit into the pipelines to which the wells may be connected. Lessee shall have the continuing right to purchase such production at the wellhead market price then prevailing in the same field (or if there is no such price then prevailing in the same field, then the nearest field in which there is such a prevailing price) for production of similar grade and gravity. Lessee may sell any royalty oil in its possession and pay Lessor the price received by Lessee for such oil computed at the well; (b) For gas (including casing-head gas) and all other substances covered hereby (i) used off the leased premises or used in the manufacture of gasoline or other products, the market value at the well of one-eighth (1/8) of the gas so used, or (ii) if sold on or off the leased premises, one-eighth (1/8) of the amount realized from such sale, provided the amount realized from the sale of gas on or off the leased premises shall be the price established by the Gas Sales Contract entered into in good faith by Lessee and gas purchaser, provided that on gas sold by Lessee the market value shall not exceed the amount received by Lessee for such gas computed at the mouth of the well; (c) If a well on the leased premises or lands pooled therewith is capable of producing oil or gas or any other substance covered hereby but such well is either shut-in or production therefrom is not being sold or purchased by Lessee or royalties on production therefrom are not otherwise being paid to Lessor, and if this lease is not otherwise maintained in effect, such well shall nevertheless be considered as though it were producing for the purpose of maintaining this lease, whether during or after the primary term, and Lessee shall tender a shut-in payment of One Dollar per acre then covered by this lease, such payment to be made to Lessor or to Lessor's credit in the depository designated above, on or before 90 days after the next ensuing anniversary date of this lease, and thereafter on or before each anniversary date hereof while the well is shut-in or production therefrom is not being sold or purchased by Lessee or royalties on production therefrom are not otherwise being paid to Lessor. This lease shall remain in force so long as such well is capable of producing and Lessee's failure to properly pay shut-in payment shall render Lessee liable for the amount due but shall not operate to terminate this lease. The intermittent

Fig. 4–1. Typical oil and gas lease. There is no such thing as a "Standard Lease."

Damages

After the lease has been negotiated, executed, and recorded, the lessee (the operator) may begin to explore. This usually involves clearing paths for seismic surveys, constructing roads and drill sites, and eventually building facility sites and installing pipelines. In all cases, the operator will negotiate a damage settlement with the surface owners (who might be different from the mineral owners). The latter cannot, according to the law in most states, refuse access to the site where the minerals beneath the ground can be tapped. But they are entitled to compensation for any damage to their property or crops. Since the United States is heavily explored and drilled, damage amounts are normally set according to prevailing customs established in the area.

Since the surface owners may not have an interest in the minerals and their potential revenue, they may not want anything to affect the aesthetic value of their property. That can become a contentious issue. Operators must always obtain permits to drill from the state. Property owners without mineral interests sometimes come together in an alliance to attempt to stall or prevent drilling and production operations, through hearings or other political persuasion. This activity is particularly prevalent in the western United States and Canada, where vast tracts of land with severed mineral rights overlie coal bed methane reserves.

Non-U.S. Rights

Gaining access in non-U.S. countries, where the mineral rights are owned by the governments, requires competitive bidding or negotiation by the interested companies. Almost always, they must commit to an investment in any parcel in which they have an interest. Their interest may have come from interpretation of a few seismic surveys they ran themselves or acquired from a predecessor; it may have come from geologic interpretation of one or more unsuccessful wildcats in the area. In most cases, the promised investment amounts to several million dollars or more and may include drilling one or more wells. Investment of this size is generally beyond the means of individuals. The bidding is usually the province of corporate entities or consortiums.

Bidding Process

Prior to bidding, for federal or state properties in the United States and properties in other countries, the government will have divided its onshore or offshore area into blocks of land, delineated by metes and bounds on land and by degrees of latitude and longitude at sea. In the offshore Gulf of Mexico, for example, the blocks are normally 5,000 acres. In the North Sea, they range from 100,000 to 125,000 acres, depending on who owns the area—the United Kingdom, Norway, Germany, Denmark, or the Netherlands. In other oceanic areas, the lease blocks can be far larger, depending on the inclination of the owning nation. Regardless of what country has ownership, the size of the parcels can be substantial.

Every government has a favorite form of contract or operating agreement. Some governments favor a profit-sharing agreement that allows the operator to subtract portions of the investment and costs from revenue before sharing profits with the host government. Some governments require that a state-controlled company own a share of any discovery and share proportionately in the profits. Other governments call for a fixed dollar amount of return for every barrel of oil or cubic foot of gas produced.

Almost all governments with any mineral wealth have a ministry of oil and gas that oversees operators' activities. In the United States, the Department of the Interior (DOI) has jurisdiction over federal lands, both onshore and offshore. Within the DOI, the Minerals Management Service (MMS) deals with offshore federal lands, and the Bureau of Land Management (BLM) is the onshore counterpart. The MMS and the BLM orchestrate the bidding process, set the royalties, oversee operations, and otherwise look after the nation's interests. Federal lands can lie within a state (e.g., in national parks and forests) or off the coast of a state. In every case, the revenue collected from these federal lands is shared with the state. During the past two decades, the MMS and the BLM have distributed $135 billion in revenue from the offshore alone to the nation, various states, and the American Indians.

Pooling

Often, private, nongovernment owners do not own enough land for an oil or gas well to be drilled on their property. If such is the case, their land will be *pooled* (combined) with their neighbors' until enough land is held by the lessee to satisfy state rules and regulations regarding *spacing* (the amount of land assigned to each well). In the U.S. onshore, the states regulate the minimum amount of land assigned to one well.

Suppose the land is in Texas and one landowner has contributed 10 acres and another 30 acres where an investor wishes to drill. The minimum acreage assigned to a wildcat in Texas is 40 acres, known as the well's *drilling unit*. The two neighbors' interests will be pooled, and their share of the royalties from the well will be divided 25:75 in accordance with their individual ownership, regardless of whose land the well is actually drilled on.

Imagine another pooling situation—offshore in the Gulf of Mexico, where two blocks have been leased to two separate investing combines (groups of oil companies). One wildcat is drilled, and a discovery is made that seismic data clearly show to extend onto the other block. Geologic evidence and reservoir evaluation indicate that only one production platform and only a few wells are needed to drain the entire accumulation. Further, it is clear that none of the wells need be drilled beneath the so-far-untested block. Unless the ownership of the two blocks is pooled, unnecessary investment in wells and production facilities must be made on the nonproducing block to perpetuate it beyond its primary term and to protect it from drainage. The solution is to pool the interest of the two parties, who must first negotiate terms suitable to themselves and then have that agreement approved by the MMS.

Nonparticipation

Sometimes the land that is pooled contains small plots, comprising only a few percent of the total acreage in the unit. In such a case, small-interest owners may decide not to lease to the drilling company. They have two alternatives—share in the cost of drilling or refuse to participate. If they choose to refuse, the investing company (the lessee or working-interest owner) may decide to drill anyway and place the party standing out of the venture in a *nonparticipating* position. The investing company has to stand the cost of its own interest and that of the nonparticipating party. But as the law works in many states, if the well is successful, the investing company gets paid back before the nonparticipants receive any revenue. The amount of payback is equal to 200% of the nonparticipating party's theoretical share of the investment, before the nonparticipating party gets any income.

Nonconsent

In some ventures, more than one party may join as a working interest owner, bearing the costs of the investment. All operating agreements for these arrangements have a similar nonparticipation section that the working-interest owners sign. In this case, it is known as *nonconsent*. At various stages in the drilling of a well or its workover, the parties to the investment are given the opportunity to agree or disagree with the recommendations of the lead working-interest owner, the operator. If some nonoperators disagree and cannot persuade the operator to change the recommendation, those disagreeing nonoperators may elect to go nonconsent. The operator and those who consented may decide to proceed, bearing the entire future cost. If the operation is successful, the operating agreement between the parties will allow the consenting parties to recover 300%—and sometimes up to twice that—before the nonconsenting parties "back in" to receipt of income from the well. (There is some logic behind the nonconsent amount: it is high enough that the reluctant parties make a noticeable sacrifice if they stand out. At 600%, few working-interest owners are prepared to give up so much potential revenue. They often join in the investment, albeit reluctantly.)

Rewards

In both cases—where the minerals owners opt out of the project and where some of the working-interest owners opt out later on—the seemingly large penalty borne by the nonconsents is also meant to compensate the operator and the participating parties for assuming risk. After all, they are committing substantial amounts of money, and the well may produce no commercial hydrocarbons.

Some of the big decisions that trigger checking with the partners include:

- drilling deeper
- completing the well

- abandoning the well

- drilling development wells

- working over wells

Unitization

Another event that may be appropriate in the life of a lease is unitization. Primary production operations can usually be pursued without much regard to what is going on next door, on a neighbor's land. But enhanced oil recovery and waterflooding (operations discussed in chapter 8) each require a cooperative effort among all of the working-interest owners sharing in production from the entire reservoir. If conversion to this secondary phase of recovery is deemed economically attractive, then all of the lands must be pooled or unitized so that the revenues are fairly distributed among the owners.

The parties negotiate elaborate formulas that consider the amount of remaining oil under each property, prior investments, the condition of the wells and facilities, and the amount of current income, to name a few important factors. Once the parties agree to a revenue- and investment-sharing formula, each royalty-interest owner and working-interest owner signs a unit agreement. Then, each working-interest owner signs an operating agreement that designates, among other things, which company will operate the project and the fees that it will charge. Generally, the party holding the largest working-interest ownership is named the operator because it has the largest potential gain (or loss) from a properly (or poorly) conducted operation. In some states, the two agreements are combined into one document.

The second- and perhaps third-largest working-interest owners usually act as watchdogs. Normally, the operating agreement will call for (at least) annual meetings of the owners at which the operator presents a consolidated report of operations for the year, deals with any criticisms, and discusses investments or operational changes visualized for the next year.

The Texas Tidelands Controversy

Few squabbles over ownership are more bizarre than that between Texas and the federal government over the submerged lands off the coast of Texas. Texas acquired this land by establishing and maintaining itself as an independent nation until it joined the Union in 1845. Texas' rights originated in 1836 on the battlefield at San Jacinto, when it won its independence from Mexico. On December 19, 1836, the First Congress of the Republic of Texas enacted into law the boundaries of the new nation, using a sketch done by General Sam Houston at San Jacinto.

The boundary in the Gulf of Mexico was described as "beginning at the mouth of the Sabine River, and running West along the Gulf of Mexico three leagues from land" (one league equals three miles). This boundary was well known by the United States when it recognized the independence of the Republic of Texas. The Republic would not agree to annexation by the United States until assured by President James K. Polk in 1845 that the United States would "maintain the Texian title to the extent that she claims it to be." This position was supported by dozens of U.S. Supreme Court rulings. Texas leased rights to oil, shell, kelp, fish, ports, piers, docks, and expensive building sites on these lands.

Then, oil and gas were discovered off the coasts of Louisiana and California. By 1950, there were 1,031 federal lease applicants blanketing the coasts of Texas, California, and Louisiana, as the industry inched out to sea. A fact aggravating the oil companies at the time was that the price of leases on state lands was substantially higher than on any federal lands. The oil companies, not being careful of what they asked for, persuaded the federal government to attack the claims of the states, including Texas, to the mineral rights of the tidelands. The feds were easily persuaded because they lusted over the potential royalty income.

This controversy became a hot political issue in the 1950s. President Harry S. Truman vetoed two bills by the U.S. Congress recognizing state ownership of the property. In the Presidential campaign of 1952, General Dwight D. Eisenhower made special recognition of the rights of Texas, as well as the long-recognized rights of the other states as upheld under earlier Supreme Court decisions. He declared in favor of state ownership and promised to sign the bill if enacted by Congress.

The Democratic nominee, Adlai Stevenson, said that he would veto such a bill. In Texas, this became a crucial issue in the 1952 campaign.

The Texas State Democratic Convention voted to place Stevenson's name on the ticket, but then passed a resolution urging all members of the Texas Democratic Party to vote for Eisenhower. Eisenhower carried the state in the November elections.

In 1953, the U.S. Congress passed the Submerged Lands Act, granting the states the rights to their tidelands and limiting state jurisdiction to three miles from the coast, but in the case of Texas and Florida, three leagues. Beyond that belonged to the federal government. President Eisenhower quickly signed it into law.

By this act, Texas and Florida ended up with nine miles of offshore control while the other states had to be satisfied with three.

Perpetuation and Termination

An important aspect of any lease, foreign or domestic, is perpetuation of the lease beyond its primary term. That time limit may vary from 3 to 10 years. Some last 20 and even 30 years. Whatever the specific term, a well must be producing on the lease in paying quantities for the lease to be perpetuated. At the end of the primary term, or if production has ceased to be economic during the term, the lease reverts to the lessor. Other terms of the lease may trigger premature termination. A common one is passing a deadline to make a specific investment, such as drilling a well or running a seismic survey. In some states in the United States, after the primary term passes, all lands that have not proven productive revert to the lessors. This may be not only in a horizontal (surface) sense but vertical as well. For instance, if a well is producing at 8,000 feet, then any intervals below that depth (usually starting 100 feet below the deepest producing horizon) revert to the lessors, and they may lease this deeper zone to any interested party.

And On...

So far the questions of where the hydrocarbons came from, how they got there, and who owns them have been addressed. The rest of the story is about how to reach, evaluate, and produce them.

5

Getting There: Drilling

There are three ways to *make hole*, that is, to penetrate the Earth to reach a hydrocarbon resource:

- digging with a pick and shovel or, more recently, with heavy-duty ground-moving equipment

- punching a hole, using a large chisel-shaped device

- rotating a bit on a shaft, much like drilling a hole in a piece of wood

All these methods have been used at some time in the quest for oil and gas. For centuries, men dug tar and pitch from below surface seeps. Today, miners strip-mine or open pit mine to recover the heaviest of the hydrocarbons—oil sands. The second, cable tool drilling, is still being used in many specialized situations, and the third, rotary drilling, is contemporary and the most relevant. But the grandfather of the rotary drill, the spring pole, is worth a short look.

The Spring Pole

As long ago as 2,000 B.C., the Chinese punched holes in the ground to recover salt brine, using bronze bits and cane for *casing*. Most of their holes were less than 200 feet deep, but there is record of at least one hole going down 3,000 feet. Water well diggers in North America since the 16th century used a spring pole, the ancestor of the cable tool rig (fig. 5–1). With this device, they pounded a heavy iron bit into the ground. (In some early operations, they even used a slim tree trunk with a bit on the end.)

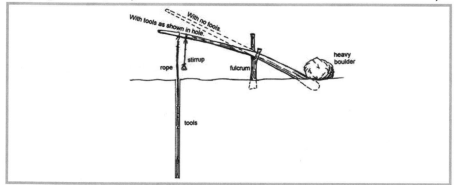

Fig. 5–1. Spring pole drilling assembly. The "driller" jumps into and out of the stirrup pulling the pole dawn and releasing it in a rhythm that allows the bit to smash into the ground. Courtesy of S.T. Pees Associates.

With exhausting effort, they pulled down the end of the pole from which the bit hung, letting the chisel smash into the bottom of the hole. Then, they released the pole, allowing it to spring upward, lifting the bit into the air, ready for another downstroke. This crude and grueling operation resulted in holes only tens of feet deep. Not until the invention of the steam engine and its application to the cable tool rig was there a way to penetrate the Earth that didn't wear out the operators as fast as the bits.

Cable Tool Drilling

Punching a hole in the ground with a cable tool rig (fig. 5–2) also requires dropping a heavy bit repeatedly with such force that the struck rock shatters. The cable tool bit is a heavy iron bar 4–10 feet long, weighing about 150 pounds, with a chisel shape at the striking end (fig. 5–3). At the top of the bar is an eye through which a cable is attached. That cable is fixed to the end of a wooden beam centered on a fulcrum, altogether a device much like a child's teeter-totter.

Fig. 5–2. Cable tool rig from the 1920s. Courtesy of Canadian Petroleum Interpretive Centre.

Fig. 5–3. Cable tool bits. Note the design has evolved from a chisel-like end to a fluted column with two chisel points at the end. The flutes allow the bit to drop through water and rock debris more easily than a solid bar. Courtesy of Barr Colony Heritage Cultural Centre—City of Llydminster.

When one end of the beam goes down, the other end rises, lifting the bit (fig. 5–4). From there, the bit is dropped to create the striking force that shatters the rock. Power comes from a steam or diesel engine that rotates a large wheel through a system of belts. The wheel's rim is attached to one end of the beam. As the wheel turns, the beam rises and falls, and the cable carrying the bit at the other end goes up and down.

After a while, so much rock is shattered that it cushions any more blows of the bit. To drill deeper, the rock fragments must be removed from the hole. The drillers pull the bit

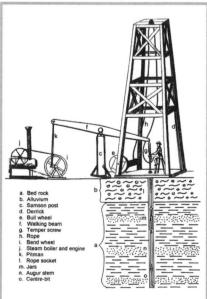

a. Bed rock
b. Alluvium
c. Samson post
d. Derrick
e. Bull wheel
f. Walking beam
g. Temper screw
h. Rope
i. Band wheel
j. Steam boiler and engine
k. Pitman
l. Rope socket
m. Jars
n. Augur stem
o. Centre-bit

Fig. 5–4. Cable tool rig schematic. Courtesy of S.T. Pees Associates.

out of the hole and drop in its place an extended, bucketlike device called a *bailer*, a hollow iron or steel tube with a flap on the bottom. As the bailer reaches the bottom of the hole, a lever opens the flap, allowing the chips to rise into the bailer. As the bailer is pulled out, the flap closes. The bailer, full of chips, is pulled out of the hole and dumped.

The top end of the bailer is attached to a cable that runs over a pulley or *crown block* at the top of the derrick. From there, the cable goes to a winch powered by a steam or diesel engine.

The derricks of the 19th and early 20th centuries were made of wood and were 30–40 feet high with a square base about 15 feet on a side. Some benevolent rig owners enclosed the sides of these derricks to protect the drillers and crew from the weather. Early in the 20th century, steel derricks replaced wooden ones and, in a less compassionate world, were left open to the elements.

Cable Tool Operations

As described in chapter 3, the pore space below the water table is filled with liquid, which is, alas, almost always water. Above the water table, in the *zone of weathering*, pore spaces are either air filled or have moisture from rainwater moving down toward the water table. As the bit's pounding takes place, the drillers need to assure the presence of some water. For the bailer to work properly, the chips have to be in a slurry. At shallow depths, in the zone of weathering, the drillers may have to pour water into the borehole. Below the water table, usually enough water enters the borehole naturally.

Sometimes a very porous and permeable rock layer releases water with such a rush that it exceeds the capability of a bailer to keep the hole dry enough to drill ahead. In that case, the drillers drive casing (hollow pipe of various diameters through which the bit passes) into the hole to the deepest point the bit has reached. With the water held back, drilling can resume. However, if another porous rock layer is penetrated and the water inflow rate exceeds the capability of the bailing operation, then the drillers may have to run another string of casing in the hole.

Each successive casing string is run inside the previous string, so that each has a smaller diameter than its predecessor. Eventually, there is no room left for the bit, and drilling must cease—a bad outcome and a time for finger-pointing by the drillers and the geologists if the target has not been reached.

If the rig's bit happens to penetrate a rock whose pore space contains hydrocarbons, they may pour into the empty hole and gush to the surface, proclaiming for all to see that the well is a success. In a simpler, less ecologically conscious time, wildcatters delighted at the sight of a gusher. Nostalgic photos of oil shooting out of a well, up past the derrick, and into the sky make environmentalists shudder today.

Drilling a well with cable tools is a slow process. Pound up one foot of rock, pull the bit, bail out the hole, *redress* the bit (resharpen the chisel face), slip the line suspending the bit so that there is enough to allow the bit to slam into the bottom, and continue punching the ground again. In the early days of cable tool drilling, making a few feet of hole in a 12-hour shift was a goal, but hardly a standard. Fortunately, in those times, the objective hydrocarbon reservoirs were found close to the surface. Drake's 69½-foot-deep well, which started the American oil industry, took nine months to drill. Subsequently, cable tools have been used to drill as deep as 7,000 feet, but nowadays the target depth is more often 2,000–4,000 feet.

Modernization of the cable tool drilling industry took the form of mobilization. Contractors built steel masts to replace the wooden derricks and put the whole assembly on wheels, allowing it to move from one site to another in a day. Today, cable tooling is often used for reentering and reworking existing wells or for simple completions. Cable tool rigs are inexpensive to operate, but they can be employed only where the operation is uncomplicated and the pressure in the reservoir starts out low or has already been depleted. For example, workover operations are routinely conducted with cable tools in the huge Panhandle Field in West Texas, where the reservoir pressure has been depleted to 50 psi or less, and the wells are less than 3,300 feet deep. Fields like this abound in the U.S. Midcontinent (fig. 5–5).

As wildcatters moved their search from the northeastern United States to the U.S. Gulf Coast, geology played a cruel joke on them. Much of the rock strata was too soft and unconsolidated to stay in place. The sides of the drilled holes quickly collapsed before casing could be put in the hole. Drillers had to find a new technique.

Fig. 5–5. Early Oklahoma field drilled with cable tools. Horse and rider in the foreground and the absence of any motorized vehicles dates the picture.

Rotary Drilling

On October 27, 1900, Captain Anthony Lucas, an Australian-born mining engineer, started drilling atop a small prominence near Beaumont, Texas, called Spindletop Hill. He had hired two water well drillers from East Texas who had been using a rotary drilling rig. According to the present curator of the Spindletop Museum, rotary drilling had never been used successfully in the oil and gas industry before then. On January 10, 1901, from a depth of 1,000 feet, the well abruptly gushed 80,000 barrels per day into the air. The average production rate of oil in the United States the previous year had been only 174,000 barrels per day. Spindletop was no small event.

This discovery heralded two important changes in the industry—the effectiveness of rotary drilling and the association of oil and gas with *salt domes*. Spindletop Hill is the geological result of an upwelling of a large, underground salt deposit creating a salt dome. Over the previous tens of millions of years, as the salt surged upward (perhaps *surge* is an overstatement), it formed seals against the porous rock that it pushed through. Later, these sealed layers became hydrocarbon traps. Many of the strata were not completely sealed, as evidenced by the oil and gas seeps around the hill and in the nearby, appropriately named Sour Lake.

But back to the first part of the good news, rotary drilling. At first glance, the main difference between rotary drilling and cable tool drilling appears to be the motion. Instead of punching a hole by pulverizing the rock, in rotary drilling the bit is spun (rotated) to break up the rock. But just as important was the use of *drilling mud* (a slurry of water and clay), which among other things was the method for removing the chips. In rotary drilling, the bottom-hole debris is removed by circulating fluid down the drill pipe, around the drill bit, and back to the surface outside the drill pipe, carrying the rock chips with it.

At Spindletop, clear water would not keep the rock debris in suspension long enough to get it out of the hole. So, with legendary Texas inventiveness, Captain Lucas and his crew filled a circular area near the well with water and drove cattle in a circle through the water. The result was drilling mud. The thickness of the resulting slurry was sufficient to keep the rock chips in suspension long enough for them to be circulated from the hole. Besides that, the mud had a plastering effect that prevented the sides of the hole from collapsing, and the drillers could concentrate on *making hole*, once again drilling ahead.

Rig Components

The typical rotary rig (fig. 5–6) has a host of operating parts, each performing a solo function but together operating in a mechanical symphony:

- a derrick (or mast) that provides a frame from which to raise and lower whatever needs to go in the hole, such as the drill string, casing, or wireline tools
- a hoisting system
- a mud-mixing and -circulating system including pumps and tanks
- an apparatus to grip and rotate the drill string and an engine to drive it
- blowout prevention equipment to preclude pressurized gas and liquids from escaping the well
- racks to hold drill pipe and casing
- housing to shelter meters, equipment, sacked chemicals, and dry mud and office space for the rig manager

Fig. 5–6. Rotary drilling rig. The smaller rig to the left has a capability to 12,000 ft. The other is a rig capable if drilling to 20,000 ft. or more. The more massive derrick of the larger rig supports heaver casing and drill pipe loads and stands higher above the ground to provide clearance for a larger BOP stack. Courtesy of Helmerich & Payne International.

The derrick

The derrick or mast is a steel structure up to 170 feet tall. Land derricks are constructed in such a fashion that they can be laid down and disassembled into four or five large pieces for moving, on trucks, from one drill site to another. Offshore, a derrick is a semipermanent structure mounted on a drilling vessel or fixed platform.

About 80 feet above the rig floor and attached to one side of the derrick is a work platform known as the *monkey board*, a name probably indicating how the rig workers move about on the derrick. On this platform, the derrick man works to manhandle the drill string from the top as it is removed or run into the drilled hole.

The size of the derrick, both its height and structural massiveness, is determined by the rig's mission, mostly the depth of the wells it is to drill. As drilling proceeds, the derrick must be able to suspend the entire length of the drill string or the casing. For a 1,000-foot hole, the drill string weighs about 16,500 pounds. For a 20,000-foot hole, the rig would have to support at least 350,000 pounds. Casing, with wider diameter and more steel, is even heavier than drill pipe, so derricks used for deepwater drilling are designed to suspend at least 1,000,000 pounds. Part of this strength is to support the load of the string, and part is to handle the additional strain when a string gets stuck and has to be forcibly pulled out.

Simpler structural requirements make the smaller rigs used for shallow wells cheaper to build, less expensive to use, and more mobile. For rigs drilling to deeper depths, mobility is sacrificed, and the structure is made massive enough to hold the weight of the longer strings of drill pipe and casing.

Smaller rigs can be moved from one location to another on land in 24 hours. The larger rigs may require a week or more. Offshore, of course, the mobility of the rig is a function of the vessel supporting it and not of the structure itself. Drillships can move quickly and over great distances from job to job. The jack-up and semisubmersible drilling rigs described later lumber from place to place under tow. Drilling rigs on fixed platforms are usually semipermanent fixtures moved around the platform on rails, but when they are no longer needed, they require tedious disassembly and transfer to barges to move them to another location.

The hoisting system

A rig's hoisting system is used to withdraw the drill string from the hole, to replace a dull drill bit and to add additional pipe to that same drill string as the hole is deepened. It is also used to support and lower casing when it is run in the hole.

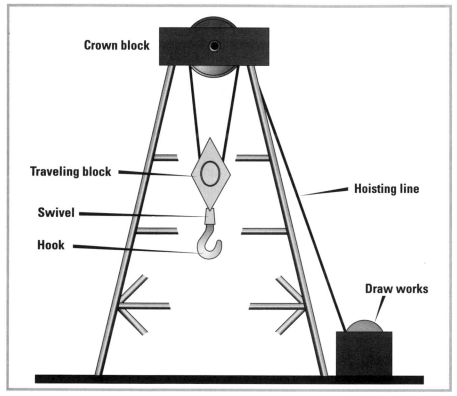

Fig. 5–7. Hoisting system on a rotary drilling rig

Major elements of the hoisting system (fig. 5–7) include the following:

- The crown block, the upper part of the pulley system, is a set of sheaves (grooved wheels) at the top of the derrick.

- The traveling block (another set of sheaves), at the bottom of the pulley system, includes the drilling hook that hangs from it.

- the swivel is a mechanical device that suspends the weight of the drill string from the hook to the *kelly* but allows the kelly to rotate. Into the swivel goes the kelly hose, through which mud circulates from the mud tanks down the kelly and then down the drill pipe.

- The elevators, clamps that go around the drill pipe and bear the load, are attached to the hook and are swung out of the way while drilling. Elevators are used when adding drill pipe, pulling the drill string from the hole, or running the string back in the hole.

- The draw works (a hoist), including the drum on which the drilling line is stored, are powered by the rig's engines that provide the energy to raise and lower the drill string or casing.

At the rig floor

The drill bit must be rotated to drill deeper. The components of the rig that accomplish this have elements named the kelly, the kelly bushing, the rotary table, and the rig engines, which provide power to turn the rotary table that rotates the drill string (fig. 5–8). The kelly connects the top of the drill string in the hole to the hoisting and mud systems. The kelly bushing is the apparatus that sits in the rotary table and grabs onto the kelly.

Fig. 5–8. Rotary table, kelly, and kelly bushing. The rotary table is the round element into which the kelly busing fits. Through the busing slides the forty-foot long kelly. In the right background is a set of slips and the top end of a joint of drill pipe, sitting in the "mouse hole." Behind the slips is the "rat hole" where the kelly is set aside during the connection of the next joint of drill pipe.

Drill pipe is round and would provide little purchase to translate the rotational forces of the rotary table. The kelly is a hollow, 40-foot-long, heavy steel pipe with a square or hexagonal cross section. It is screwed into the top of the latest piece of drill pipe to go in the hole. The kelly's shape provides a "grabability" advantage over round pipe.

Not only must the kelly provide the means by which the drill pipe is rotated, it must also move up and down through the rig floor as the hole is deepened. A smooth, vertical motion is enabled by wheels located in the kelly bushing.

The kelly bushing is firmly attached to the rotary table during drilling but is pulled out of the way along with the kelly before any drill pipe is taken out of the hole.

Drill bits

To make hole, the drill string is rotated and the bit gouges or pulverizes the rock. Penetration of soft formations is by gouging and jetting. In harder, brittle rock, progress is generally slower (fig. 5–9) and is achieved more by pulverizing and crushing the rock. The classification of drill bits in figure 5–10 is simple enough, but it belies the uncountable variations in vendors designs within each category.

Tricone bits and *polycrystalline diamond compact* (PDC) bits are the work horses of the industry (fig. 5–11). Tricone bits have three metal cones studded with various very resistant materials placed at the bottom

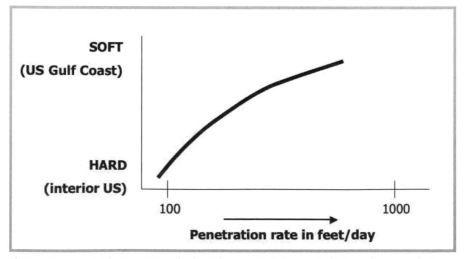

Fig. 5–9. Bit penetration rates. Graphed are the penetration rates in feet per day when drilling hard and soft rocks typical of the interior U.S. and the Gulf Coast regions.

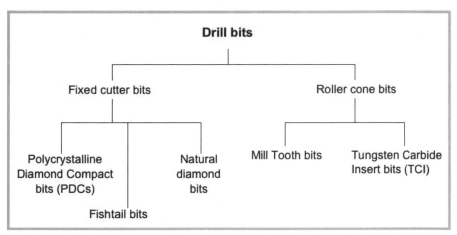

Fig. 5–10. Types of drill bits

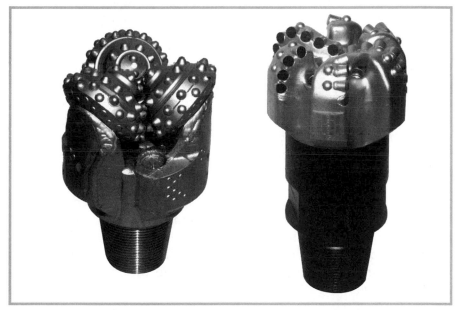

Fig. 5–11. Tricone and PDC bits. Tricone bits role across the rock pulverizing it while PDC bits have a gouging action.

of the bit. They are configured in such a way that when the drill string is rotated, they are dragged across the exposed rock. Each cone rotates, even as the bit turns, extending the life of the studs on the cutting surface. The studs are typically made of tungsten carbide or industrial diamonds, designed to withstand the relentlessly punishing forces of weight and rotation.

In soft formations such as are found on the U.S. Gulf Coast and many offshore areas of the world, the sediments have not become sufficiently compacted and cemented to make them as hard to drill through as in most inland areas. In such areas, bit design is dramatically different. Rather than crush the rock, the soft sediments can be gouged from the bottom of the hole. This type of tricone bit has steel teeth, rather than tungsten carbide or diamond studs. In very soft formations, the jets around the cones remove as much or more of the soft material as do the cones themselves.

Wear is the curse of the drill bit, shortening its life and increasing the frequency of replacement, a time consuming and costly exercise. A modern solution has been the development of the PDC bit, which has a man-made diamond material pasted to the face of each cutting edge. There are no rotating cones with their bearings to wear out, and only one PDC bit may be needed to drill an entire section of a well, perhaps as much as 3,000–5,000 feet. Around 10 tricone bits might have been required otherwise.

Penetration. Penetration rate (fig. 5–9) depends on many factors besides what is being drilled (the lithology). Weight on the bit and its speed of rotation are the next-most-important determining elements. Weight on the bit comes from the drill string. However, as the bit goes deeper, the bit cannot stand the full weight of the drill string. Drillers continuously and carefully monitor how much weight they are allowing the hoisting system to release onto the bit. They also manage the speed with which the bit turns by controlling the revolutions of the rotary table.

Changing the bit. When a bit is worn and needs to be changed, the crew has to pull the entire drill string from the hole, replace the bit, and rerun the pipe and new bit. This operation is known as *making a trip*. While very routine, it is the most time consuming and one of the toughest jobs the rig crew is required to do. To make a trip, the driller and his crew must do the following:

- Stop the rotary table by disconnecting the motors.

- Turn off the mud pumps to halt mud circulation.

- Raise the entire string of pipe until the top of the last connected joint of drill pipe is a few feet above the rotary table.

- Suspend the drill pipe with *slips*, steel wedges tied side to side, forming a near circle with steel teeth on the inner side of the circle (fig. 5–12). The slips are placed around the drill pipe as it is lowered a foot or so into the hole. The wedges fit against the sides of the opening in the rotary table, simultaneously compressing the teeth against the drill pipe, and locking everything in place. That leaves the entire weight of the drill string now suspended from the rotary table rather than from the hoisting system.

- Unscrew the kelly and swivel from the drill pipe and stand them aside.

- Lower the elevators (hanging from the hook) and latch them around the top of the drill pipe.

- Raise the drill pipe until three joints (90 feet; called a *stand*) are above the rotary table. The slips are removed and set aside simultaneously with the initial upward movement of the pipe.

- Set the slips once again around the drill pipe. Unscrew and stand the drill pipe back in the derrick.

- Repeat the operation until the bit is reached, unscrewed, and replaced.

- Reverse the entire operation and commence drilling once again.

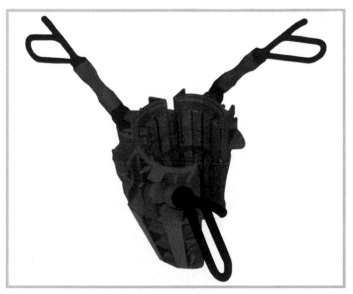

Fig. 5–12. Slips. Also shown in figure 5–8.

Making a trip from a depth of 10,000 feet may take as long as 10 or 12 hours.

Drilling Mud

The mud-mixing and -circulating system is one of the most important components of the rig. Many well problems originate in a poorly formulated drilling fluid or start with downhole circulation difficulties.

The purpose of mud

Consider the functions of mud as the bit makes hole:

- contain the Earth's pressures

- lubricate the bit

- cool the bit

- remove the cuttings

- prevent the hole from collapsing

So what is this inelegantly named substance, mud? These days, mud has evolved a few steps beyond the ooze that Captain Lucas scooped out of the cow paddy. Today drilling mud is a mixture of

- water (or saltwater offshore)

- bentonite clay (for viscosity)

- barite (for weight)

- chemical additives (for lots of purposes)

As wells were drilled deeper, removing the bit cuttings from the hole became impossible using nothing more than dirt and water. The *viscosity* or thickness of the mud became of paramount importance. A more viscous slurry keeps the rock particles in suspension long enough to allow them to be circulated from the hole. That is the function of the bentonite. Mixed with water in the rig's mud tanks, it swells and provides the viscosity necessary to suspend the bit cuttings and other materials mixed into the mud. Bentonite is a volcanic ash deposited long ago in beds in Wyoming, Georgia, South Africa, and elsewhere. It is mined, dried, pulverized, packaged in 100-pound sacks, and ultimately delivered to the drill site by 18-wheelers onshore and by supply boats offshore.

Barite, a form of barium sulfate, is likewise mined, but from deposits in limestone or sandstone formations, mainly in China, the United States, and India. Adding barite to the mud increases its density (adds weight), which permits it to hold back formation pressures. Barite has a specific gravity of 4.5, meaning it weighs about four and a half times as much as water. But why the extra weight?

As the drill bit reaches farther into the Earth, it encounters higher pressures. They can come from the weight of the water in the pores of the rock (hydrostatic pressure, described in chapter 3), from the weight of the rock itself (overburden pressure, also from chapter 3), or from a combination of the two (fig. 5–13). The pressure by water alone increases by about 0.45 psi for every foot of depth. At 10,000 feet, the pressure from water would be 4,500 psi. Other geologic forces could push the pressures higher. The driller carefully watches the flow of mud as it circulates out of the hole and into the system as a way to determine if the mud weight in the hole is sufficient to contain the pressurized downhole fluids. If not, the driller increases the weight of the mud by adding more barite.

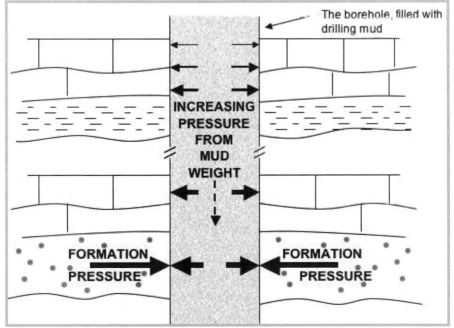

Fig. 5–13. Mud and Earth pressures. The weight (hydrostatic pressure) of the circulating fluid (mud) is intended to counterbalance the pressure contained within the formation. If it does not (is too light), fluid from the formation can invade the borehole and cause a blowout.

The chemistry of mud is complex, and choosing the right composition often requires a specialist—particularly when choosing additives to improve viscosity, reduce filtrate loss, improve bit lubricity, prevent corrosion, reduce foaming, and deal with dozens of other potential problems.

For example, as the well is drilled, some water is lost from the mud system into the porous rock through which the bit travels and is called *filtrate* at that point. Loss of water (filtrate loss) can leave the mud too thick for proper circulation. To control this problem, chemicals are added to thicken and toughen the mud residue, or *filter cake*, left on the formation face, thereby minimizing the filtrate loss.

Certain circumstances call for a very-light-weight mud system, for example where the pressure in the formation is low, often well below that of the hydrostatic gradient. In that case, crude oil, because it is lighter than water, is used as the carrying agent instead of water. Crude oil is flammable and messy and not the favorite mud system of a rig's crew, but they do what they must. Another, even-lighter-weight technique is to use air or a mist of air and water as the circulating fluid.

All drillers have experience with "special" additives to the mud system. Sometimes a well penetrates a zone where the porosity is so high or the formation's pressure is so low that the mud escapes rapidly into the formation, rather than continuing to circulate out of the hole. This is a *zone of lost circulation*. Not only is the loss of mud a concern, but a blowout may occur if drilling continues. In the history of the industry, many improbable materials have been tried to plug up the formation holes to regain circulation. In the earliest days, straw, bark chips and cotton hulls were added. In California, it was common to add cut-up Hollywood filmstrips, which acquired the name "star dust." In Texas and Louisiana, walnut and pecan hulls have been used. Even today, drillers resort to chopped hay, gravel, and whatever else is handy to deal with lost circulation.

The circulating system

The mud-mixing and -circulating system (fig. 5–14) begins in a large tank beside the rig. Onshore, fresh water is added to the tank from a nearby water well or lake or is trucked into the location. Offshore, the makeup water is taken from the ocean. The mud materials are added directly to the water, and the entire system is constantly stirred to prevent weight material from settling out of the system.

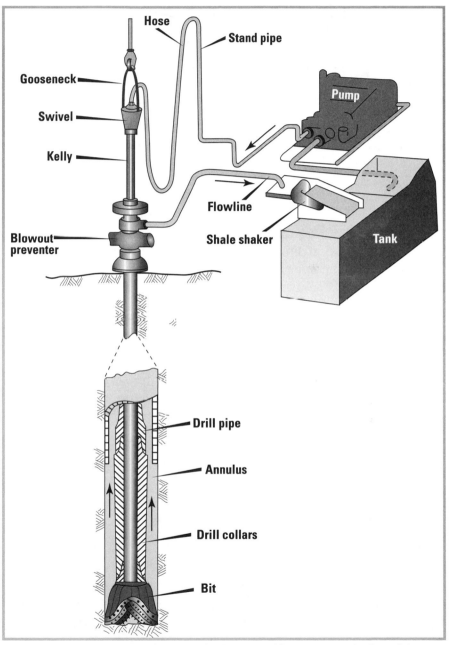

Fig. 5–14. Mud-circulation system. Drilling fluid is pumped from mixing tanks through hoses into the swivel, down the kelly and drill pipe, out the bit's jets, up the space between the drill pipe and the borehole, out of the hole, across the shale shakers, and finally back into the mixing tanks.

From the tanks, the mud is pumped up the standpipe, down the kelly hose, through the swivel into the kelly, down through the drill string, and out the jets on the drill bit at the bottom of the hole. There, the drill cuttings are picked up and circulated up the annular space (the *annulus*) between the drill pipe and the hole and, farther up, between the drill pipe and whatever casing has been placed in the hole in earlier operations.

After leaving the hole, the mud is piped back toward the circulating tanks, but first across a series of vibrating screens called *shale shakers* (fig. 5–15). These are screens set on a slope and kept in a jerky sideways motion as the mud slurry courses over them. The liquid mud passes through the screens into the mixing tank. The insoluble rock chips that do not pass through the screens fall off into an earthen pit or a tank. Onshore, some farmers like to have this residue plowed into the fields surrounding the rig. If it contains too many chemical additives, it may be hauled to certified dump sites. Offshore, the cuttings are either returned to the sea or transported to shore to the certified sites.

Fig. 5–15. Shale shakers. Drilling fluid carrying rock debris is circulated across vibrating screens. This separates the chips from the liquid mud, allowing its reuse. Samples of the chips (known as "cuttings") maybe collected from the shakers. Courtesy of Bill Booker.

Large pumps powered by equally large motors approaching 2,000 horsepower drive the mud-circulating system. Typically, the system has at least two pumps, each with its own motor located conveniently beside the circulating tank. One of the pumps operates continuously during drilling, while the second is on standby.

No one ever expects the mud system to be a guaranteed method to contain formation pressures. If circulation is lost, the remaining mud in the wellbore may not exert sufficient pressure to hold the formation fluids in the formation. The consequence could be a blowout, a rush of fluid into the wellbore and up the hole—and if not contained, out of the well. Many early drillers learned the hard way that they needed some fail-safe mechanism to prevent blowouts from endangering the lives and limbs of the drilling crews and the integrity of their equipment.

Blowout Preventers

Blowout preventers (BOPs) come in a variety of shapes and sizes, mostly dictated by the pressures they are expected to handle and the environment in which they are designed to work, particularly onshore versus offshore (figs. 5–16 and 5–17). In simple terms, they are powerful hydraulic rams that either close around the drill pipe, close against the drill pipe and pinch it together, or close off the open hole if the fluid surge occurs when there is no pipe in the hole.

Each type of preventer is a separate component of a *BOP stack*. Onshore, the BOP stack sits under the rig floor, bolted to the top of the surface casing. Offshore, the BOP stack is either at the bottom of the ocean at the *mud line* or above the water on the drilling rig platform. If the drilling rig is a floater, there are disconnecting devices at the seafloor that close off the hole when the riser between the drillship and the BOP stack is disconnected and the rig is shifted off location, as when a violent storm is forecasted.

Fig. 5–16. BOP stack ready for installation. The device at the top is an "annular" preventer used to close against the kelly. The three devices below it are used to close against the drill pipe or to shear the drill pipe to seal it. Courtesy of Helmerich & Payne.

Fig. 5–17. BOP elements. Technicians servicing the ram elements made of compressible material subject to wear. These critical parts require periodic inspection and replacement. Courtesy of Helmerich & Payne.

Other Features

Casing and drill pipe are stored on *racks* awaiting use (fig. 5–18). Casing is made in lengths ranging from 25 to 45 feet; the drill pipe usually comes in 30-foot joints. Casing and drill pipe are both added one joint at a time to the string of pipe already in the hole. Casing, on the one hand is taken out of the hole in most cases only when the well is abandoned. (In many onshore locations, abandonment regulations require that most of the casing strings be pulled and replaced by strategically located cement plugs.) Drill pipe, on the other hand, comes out of the hole every time the drill bit needs changing or when some activity is conducted that requires an open hole, such as wireline logging. The drill pipe is withdrawn in *stands*, three joints or 90 feet of drill pipe at a time. Rather than being placed back on the horizontal racks, the stands are *stood back* in the drilling rig's derrick on end, vertically.

In an onshore drilling operation, management personnel work in trailers containing desks, communication equipment such as two-way radios and telephones, instruments monitoring the condition of the well, and often beds and eating facilities. Also on location are buildings and metal sheds sheltering motors, generators, sacked mud and chemicals, and other incidentals necessary for the operation of the rig. Remote locations often have trailers for crew sleeping quarters and for cooking and eating. Offshore, space limitations and remoteness generally result, ironically, in far less primitive living accommodations and conditions (similar to the contrast in living conditions between the modern U.S. army and navy).

Fig. 5–18. Casing on a rack

Drilling Offshore

Drilling from a floating platform at sea—a semisubmersible floating rig or a drillship—calls for some special controls and equipment. As the drilling vessel heaves, rolls, yaws, pitches, surges, and sways—and all these different movements do happen in the mildest seas—the changing tension on the drill pipe could cause it to buckle. In addition, to point out the obvious, the spot where the drill bit starts making hole is unseen, maybe 1,000 or 10,000 feet below the rig floor.

Platforms for drilling

In the middle of the 20th century, when offshore drilling began in earnest, drillers realized that installing fixed platforms to drill wells would be too expensive and time consuming. Enter the mobile rigs—the submersible,

Fig. 5–19. Offshore mobile drilling rigs. Clockwise from the top, a semi submersible, a jack-up, and a drill ship. Courtesy of Transocean Drilling.

the jack-up, the ubiquitous semisubmersible, and the drillship (fig. 5–19). The choice depends largely on the water depth:

- submersibles and anchored barges up to 50 feet, usually in backwaters and swamps
- jack-ups in 25–400 feet
- semisubmersible in 200–5,000 feet, sometimes more
- drillships in 200–10,000 feet or more

At the seafloor

To provide guidance at the seafloor, the driller places a template on the seabed at the drill site. It has fittings and attached lines or transponders that allow the driller to place tubulars accurately while spudding the well, placing a conductor pipe, and so on. The BOP is generally placed on the template. A riser or riser pipe extends from the BOP to the drilling vessel (fig. 5–20). Some rigs have the BOP at the top of the riser, on the platform.

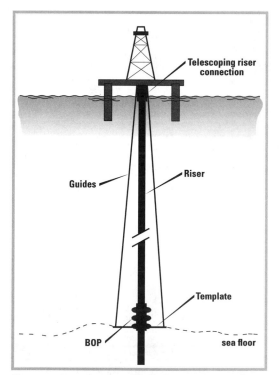

Fig. 5–20. A marine riser. The device connecting the subsea BOP or wellhead to the drilling platform vessel.

Compensators

When the drilling is done from a floating platform (a semisubmersible or a drillship), to allow for the vessel movement, the top of the riser has a telescoping joint that is attached to a heave compensator by tensioner wires (fig. 5–21). The compensator operates hydraulically, using nitrogen-pressured cylinders to keep a controlled tension supporting the riser.

The same vertical movements affect the drill string, and another system deals with that: the traveling block that suspends the drill pipe has its own hydraulic compensator to enable the driller to put the correct weight on the drill bit.

Fixed platforms

In shallower waters, fixing a mobile platform to the sea bottom is feasible—using a barge, a submersible, or a jack-up. In addition, after

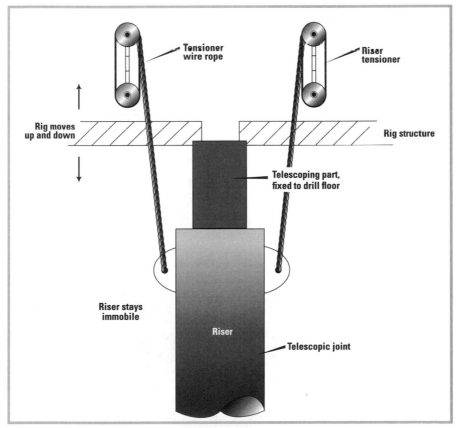

Fig. 5–21. Heave compensator. This device compensates for wave action with a telescoping segment included in the riser assembly.

permanent production platforms have been put in place, generally more development wells are drilled from it. In all these cases, the drilling procedures combine offshore and onshore techniques. Heave compensators aren't necessary, but templates and riser pipes are.

Top Drives and Automation

On some newer drilling rigs—particularly offshore rigs and drillships used in deeper waters, where time is everything—*top drives* and automated tubular handling devices have replaced the classic rotary table, kelly bushing, kelly, and swivel (fig. 5–22). Drill pipe is grabbed from above by the top drive system, which slides vertically on a track inside the derrick as the pipe enters the well. Along with introduction of top drive, the manhandling of the tubulars—drill pipe or casing—is replaced by robotics.

Fig. 5–22. Rig equipped with a top drive drilling assembly. The top drive system for rotating the drill pipe and bit is an alternative to the rotary table, kelly bushing, kelly, and swivel. Courtesy of Helmerich & Payne.

The Course

To enable the bit to drill a nearly vertical hole, a half dozen or more *drill collars* are placed just above the bit. These collars weigh 30–200 pounds per foot versus the 16.5 pounds per foot weight of normal drill pipe. With this heavy weight at the bottom, the entire string can act like a pendulum or plumb bob. Whenever it veers from vertical, gravity tends to pull it back.

Still, the bit tends to create a corkscrew path under the drilling rig. The course the bit takes is affected by the different types of rock and the dip (slope) of the rock formation. A bit tends to drill perpendicularly into the dip of a rock formation, that is, into whatever is the angle of the plane of the formation. If the formation is sloping, the path of the bit will trend away from vertical.

Before the well is drilled, the geologist has agreed with the driller that the well will end up within a specified distance from a targeted bottom-hole location. A reasonable distance could be as much as 200 feet. If it becomes apparent that the well is going to end up too far outside the agreed-upon target area, the driller can set a *whipstock* (fig. 5–23) and divert the well back in the proper direction.

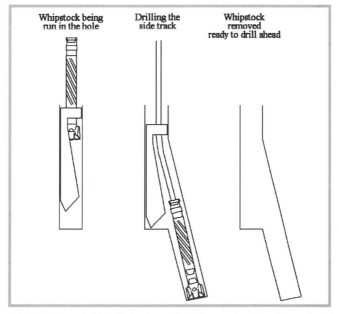

Fig. 5–23. Whipstock. This device is used to change the direction and angle of drilling. It is set at the bottom of the hole, oriented in the desired direction, and used to divert the bit from vertical to a lesser angle.

Sometimes the target is a spot thousands of feet—or even miles—laterally away from the drill site. For example, an oil reservoir can be underneath a lake, beneath a mountain range, or near the shore but still under the ocean. Drilling a *directional well* may satisfy both the driller's cost concerns and the geologist's need to penetrate a specified target (fig. 5–24). Offshore, production wells are often drilled from a fixed platform to bottom-hole locations that may be thousands of feet away laterally. Under these circumstances, a driller will intentionally divert the drill bit away from vertical by setting a number of whipstocks. The ultimate diversion, a horizontal well, is drilled sideways through a reservoir to accelerate the rate at which the hydrocarbons can be evacuated and increase their total recovery. A horizontal well, as well as many highly deviated wells, is achieved by placing at the bottom of the drill string a motor energized by the mud circulated through it. The downhole motor, when equipped with a short piece of drill pipe called a *bent sub* (fig. 5–25) can move the bit in any direction specified by the driller from the surface.

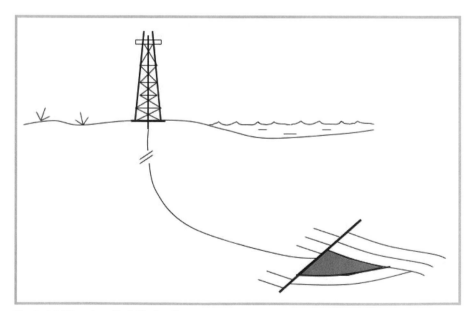

Fig. 5–24. Directionally drilled well

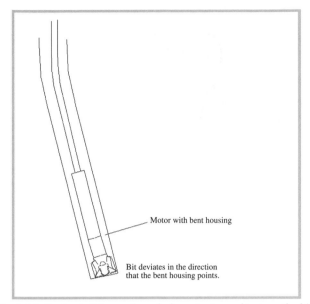

Motor with bent housing

Bit deviates in the direction
that the bent housing points.

Fig. 5–25. Bent sub. When a down hole motor is used to turn the bit, its direction is changed by adding a bent piece of drill pipe (rather than a whipstock).

Location, Location, Location

The bit has finally penetrated the rock strata predicted by the geologist, and the driller has therefore reached *total depth* (TD). The driller knows how far he has drilled by measuring the length of each joint of drill pipe as it was added to the string. But because of the corkscrewing of the hole as it went deeper, the driller does not know how far below the surface the bit has reached (fig. 5–26). The geologist knows the *measured depth* but not the *true vertical depth* of the well. Maps made by the geologist depend on the correlation of specific rock strata from various control points in terms of their depth below or above sea level (most often below, then called the *subsea depth*). Therefore, to compare where the well is at TD with the predictions, the geologist has to calculate the TD in terms of its true vertical depth (TVD).

In addition to the well depth, the geologist needs to know where the TD is in relation to the well's starting point. Before the well is started (*spudded*), the surface location is surveyed; then, as the well is deepened a single shot survey can be used to measure the well's inclination (deviation from vertical) and azimuth (the horizontal deviation or compass direction).

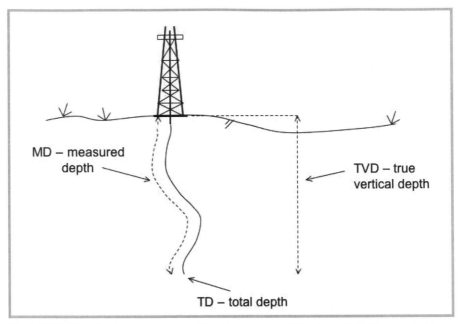

Fig. 5–26. Drilling depth measurements. The distance the well has been drilled is reported in terms of actual length that has been drilled and the depth below the surface (usually below sea level). The TD is whatever depth the well has reached, either measured total depth or true vertical depth.

Normally, these survey points are made about every 100 feet down the well. (When drilling horizontal wells and other wells where the position of the bit at every moment is critical to the success of the well, a survey can be made continuously.) All these data are combined to generate a map of the well's course. If all went well, the driller has gratified the geologist by tapping the alleged reservoir being sought.

But what has the bit tapped into? Old hands have a saying, "You're never sure about a reservoir until you test it with a drill bit." So, now that the drill bit has tapped the reservoir, their question becomes, "What have we found?"

6

What Have We Found?
Logging, Testing, and Completing

Choose thou whatever suits the line.

—Samuel T. Coleridge (1772–1834), *Names*

Once the hole has been drilled to the total depth required by the geologists, they must find out what the drill bit has actually penetrated. Have their original hypotheses held up? Do they have a producer or a dry hole? The questions have been the same since the earliest days of the industry, at least after it became environmentally, if not economically, unacceptable to prove the merit of the well by permitting it to blow out, even in the semicontrolled situation shown in figure 6–1.

Fig. 6–1. Controlled blowout at Sour Lake, Texas. Courtesy of Spindletop Museum.

Logging

Until the 1920s, the only way to tell what rock strata had been penetrated was to look at the bit cuttings that were either bailed during cable tool drilling or circulated out of the hole while rotary drilling. These rock chips were collected, inventoried, washed and dried, examined under a microscope, and documented. The physical characteristics—color, rock type (shale, sandstone, limestone, etc.), grain size, existence of porosity, and any other properties that might distinguish the chips—were recorded in a continuous fashion next to the depth the bit had reached when the chips were produced. This record of the penetration history of the well became known as a *sample* or a *strip log.*

Mud logs

The urgent need to know what the drill bit was penetrating led to analysis of the mud literally as it was circulated from the hole. Instruments were developed to detect the presence of hydrocarbons in the mud as it circulated to the surface, carrying the rock cuttings. Whenever oil or gas was detected, its presence was (and still is) called *a show*. The record of the rock chips' lithology, combined with a continuous description of the presence or absence of hydrocarbons, the rate of penetration of the bit, and the depth became known as a *mud log* (fig. 6–2).

There are two principle uses for mud log data:

- The immediate interpretation of what the drill bit has penetrated and whether there are any hydrocarbons present, an important factor that gives a qualitative indication of success or failure of the venture.

- Making maps of the subsurface geology. From the mud log (and also from correlation logs [described later in this chapter]), the geologist can find where the well lies in relationship to the surrounding rock strata found in other wells in the vicinity. For the initial well, it starts a data set to help exploit whatever may have been discovered.

A good geologist has to have the type of brain that can visualize unseen, buried, three-dimensional images, particularly in thinking about the subsurface. Contained on figure 6–3 are examples of how a geologist might map a certain subsurface layer, say a red clay. With these maps, the

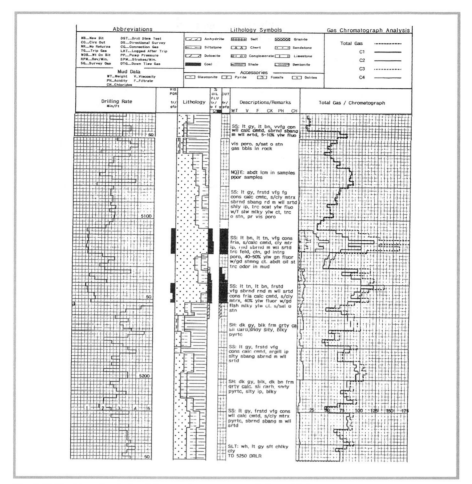

Fig. 6–2. Mud log. Analysis of the cuttings are recorded on this log.

geologist can show the amount and form of distortion that the Earth's rock layers have experienced over geologic time at this particular location.

The upper map (fig. 6–3) is a geologic subsurface contour map, which can be likened to a buried topographic map of the surface (top) of a particular rock stratum, in this case the red clay. On it are shown not only the geographic locations of wells A, B, and C but also the depth at which each well encountered the top of the red clay. Using these data (the locations and depths, collectively called *control points* or *control*), the geologist makes a subsurface map of the red clay. Contour lines are drawn evenly spaced between the control, producing a map that among other things discloses the slope down the surface from well B to well C.

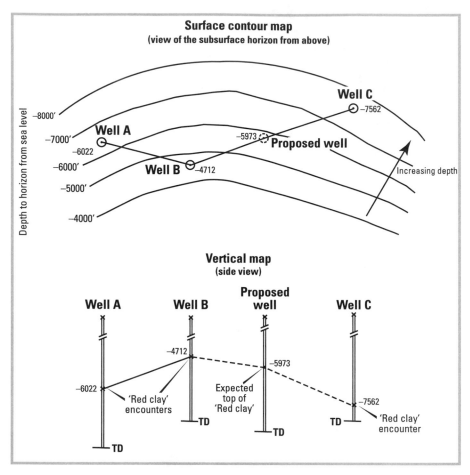

Fig. 6–3. Subsurface map combined with geologic cross section

A fourth well is proposed and placed at its proper location on the map. If drilled at that location and if the geologist's interpretation of the relationship between the data points is correct, the well will encounter the red clay at 5,973 feet

On the lower map (fig. 6-3) is another view of the subsurface. It is developed from the control points and in this instance is a side view or, in the lingo of geologists, a *cross section* or an *x-section*. While it is two dimensional, the cross section runs on lines between the well locations and is not on a flat plane. It is a flattened composite of several planes. This nonlinear positioning can cause some distortion, making some lines representing the surface seem steeper (or shallower) than they are in reality. Geologists quickly become accustomed to this phenomenon, and it bothers them not at all.

In many locales, especially along the Gulf Coast, the sands (and the shales) are indistinguishable. Geologists can tell what is sand and what is shale, but they cannot see any characteristics that distinguish between a bag of sand cuttings that comes from 10,000 feet and one from 5,000 feet. To make a geologic map of the subsurface, geologists must know how to compare what has been penetrated in one well with what has been penetrated in a neighboring well.

Fortunately, sand and shale sequences contain not only clastic (rock) material but also the shells of the minute organisms living in the seas at the time the layer was formed. These little critters, known as *foraminifera* or *forams* (fig. 6–4), were caught as sediments settled on the ocean

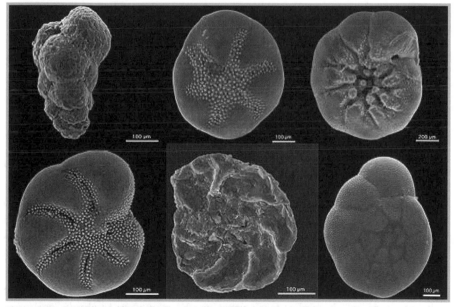

Fig. 6–4. Foraminifera. Also known as forams, each is about a millimeter across. Courtesy of USGS.

floor. The study of the shells of these organisms belongs to the science of paleontology.

From the study of innumerable samples, it has become apparent that forams developed unique shell forms depending on whether they lived in shallow, moderate, deep, or very deep water. It also has become clear that younger forams differ from older forams in consistent manners. In short, forams evolved over relatively short (in geologic terms) time periods. By cataloging the shapes, designs, and colors of these minute shells, paleontologists are able to differentiate one sand horizon (layer) from another. This gives them the common data points they can use to make geologic maps.

Knowing what rock types lie below the ground's surface has its scientific appeal. But a well is drilled to find hydrocarbons, not rock. What the rock contains is more important than what it is, at least from a commercial perspective.

So what did the drill bit penetrate? Mud logs are an indication of contents, but they supply qualitative, not quantitative, data. Two questions of paramount importance remain:

- Are there any producible hydrocarbons present?

- If so, how much?

Not so crucial but still important are a better determination of lithology and a consistent method to portray rock characteristics to enhance correlation from one well to another. And that calls for more sophisticated logs.

Open-hole logs

Conrad and Marcel Schlumberger (fig. 6–5) revolutionized the oil industry in 1927 when they produced a log of a 1,600-foot-deep well that showed the electric resistivity of the rock layers from the bottom to the top of the well. The resistivity of each rock layer depends on the amount and salinity of the water it contains. The more water there is, the less resistive the rock is to the passage of an electric current. The lower the salinity is, the higher is the resistivity. Hydrocarbons are much more resistant to the flow of electricity than is water. And freshwater is much more resistant than saltwater—so much so that freshwater can be mistaken for hydrocarbons.

The Schlumberger brothers recorded their resistivity measurements by using electric wire lines spooled into the hole (fig. 6–6). The insulated wires with exposed tips sat in drilling mud in the borehole. Even though this mud usually has a water base, it has lower conductivity than sands with salty water in them. Electric current preferentially flows out the borehole into the water that resides in the adjacent sands, then back through the borehole, through the mud to the next exposed wire tip. As the wires were

Fig. 6–5. Early experiments measuring resistivity by the Schlumberger brothers. Courtesy of Schlumberger.

drawn up the hole, the change in the resistance between the bottom wire and the upper two, recorded at the surface, showed the resistivity changes of the formation (fig. 6–6).

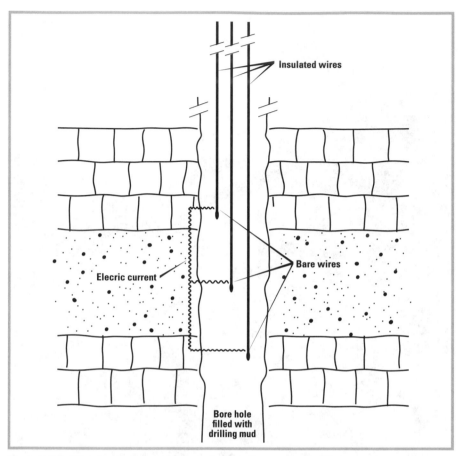

Fig. 6–6. Primitive three-wire assembly used to measure rock resistivity

The Schlumberger brothers captured these meticulous electric measurements by stopping each meter as the wirelines were pulled from the hole. Ever since, electric (and other logs) measuring downhole rock properties have been called collectively *wireline logs* or, more specifically, *E-logs*, the latter designation used for the multiple types of resistivity logs.

The objective of the Schlumbergers' logging was to determine whether oil or water was contained in the pore spaces of the rock that had been drilled and in what amounts. Their earliest attempts showed only intervals of high resistivity (possibly oil or gas) versus low resistivity (probably saltwater). But it was a beginning.

Fig. 6–7. A sonde, the housing for devices measuring rock properties in the wellbore. The technician is positioning the sonde so that it will pass through the BOPs and into the well.

Modern wireline logging is more sophisticated, of course, than what the Schlumberger brothers did. Rather than relying on bare wires, the instrumentation is enclosed in a tubular steel case, some 20–40 feet long, called a *sonde* (fig. 6–7). The sonde is suspended by a wireline (a cable) containing the electric wires that transmit the recorded properties of the formations to the surface (fig. 6–8). The sonde and the wireline are spooled into and out of the well from a logging truck (fig. 6–9). The electric impulses transmitted by the wireline are decoded by computers housed in

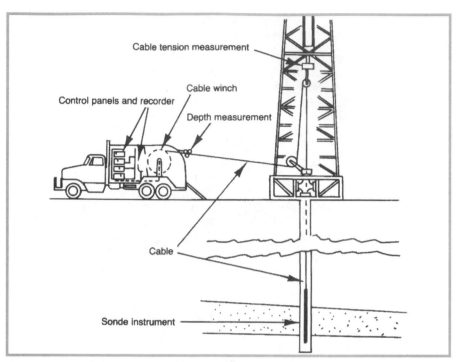

Fig. 6–8. Setup to log a well. The sonde is lowered on a wireline cable spooled from the logging truck.

the truck or, offshore, in a self-contained unit on the drilling rig (fig. 6–10). The recordings from the subsurface are printed for immediate reading and interpretation on paper records called *logs*. They are also stored electronically and often, especially in offshore operations, transmitted by either radio or satellite to the operating company's onshore office.

Fig. 6–9. Logging truck. The vehicles transports all the necessary equipment and houses the computerized recording devices.

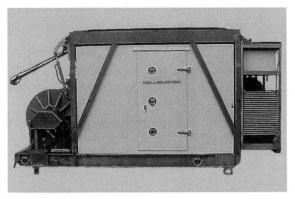

Fig. 6–10. Offshore logging unit

The early 1940s witnessed a quantum leap in the technology of well logging. Many different types of resistivity logs were invented, each focusing on a different characteristic of the rock strata penetrated (table 6–1). Some showed how deeply the mud filtrate had penetrated, giving an indication of permeability; some logs told how far the hydrocarbons had been flushed laterally away from the wellbore, indicating porosity and permeability.

Table 6–1. Resistivity tools

Type	Application
Electric log	Freshwater mud, thick beds
Induction log	Freshwater mud, air mud, oil mud
Dual induction log	Freshwater mud
Laterolog	Salt mud, highly resistive rock
Dual Laterolog	Salt mud, highly resistive rock
Microlog	Freshwater mud
Microlaterolog	Salt mud or freshwater mud
Microspherically Focused log	Salt mud or freshwater mud

The proliferation of resistivity logging techniques and tools developed for several reasons:

- Not all tools worked in all of the different mud systems used to drill the well.

- Formation conditions affected each tool's readings differently.

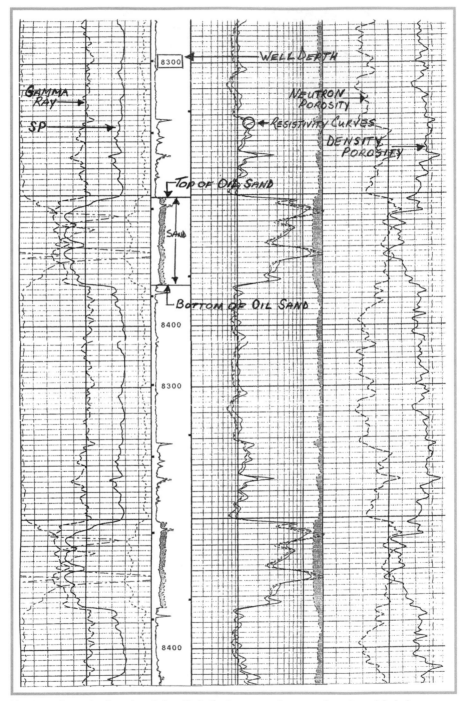

Fig. 6–11. Example of an electric log. Only the important log traces have been labeled.

Induction logs

Some formations contain clay particles that swell when contacted by water in the mud. This swelling can reduce the natural permeability of the formation to such a degree that the well may never produce near its potential. The solution is to drill using a non–water-based system, such as oil-based mud, air, or foam (a mixture of water and air). These media will not conduct electricity, so the original electric logging devices, which measure resistivity, do not work. To deal with this problem, geologists and engineers turned to electric induction resistivity logs (fig. 6–11). Instead of using exposed bare wires, the current is "induced" in the formation. A magnetic field is sent out from an induction sonde, causing a flow of current within the formations passed by the sonde. A magnetic field from the other end of the sonde picks up the current, allowing it to be recorded.

In many cases, even when logging in freshwater muds, the induction resistivity tool measures the true formation resistivity more accurately than the original electric logging devices. Different breeds of induction electric devices are now available to deal with the unique character of the rock formations in various parts of the world. Where the porosity is high and the formation resistivity is moderate, such as along the U.S. Gulf Coast and in California, the suite of resistivity logs differ from those used in the Rockies and U.S. Midcontinent areas, where the porosity values are generally lower.

Saturation

The goal of the resistivity and induction tools is to give a measure of the formation's *true resistivity* (R_t). However, bed thickness, mud type, shale content, porosity, and permeability all affect the resistivity readings differently. By running the appropriate resistivity tools, engineers can determine an accurate R_t. Empirical evidence has shown that with that value, they can calculate water saturation as a percent of total fluid present. From this value, oil saturation (S_o) is

$$S_o = 1 - S_w$$

where S_w is the water saturation. For example, in the determination of oil saturation, if the water saturation is 0.37, the hydrocarbon-filled pore space is $1 - 0.37$ or 0.63, equivalent to 63%.

Also, empirical data and laboratory experiments have abundantly demonstrated that if S_w is higher than 50%, the formation will normally produce all water. The water overwhelms the hydrocarbons. If S_w is lower than 50%, hydrocarbons and varying amounts of water can be produced.

Porosity. An empirical relationship between the measured resistivity of a saturated rock and its porosity was published by one of the industry's intellectual pioneers, Gus Archie, in 1942 and was further developed in the 1940s and 1950s. He found, as you might imagine, that the rock itself, as well as the saturating fluids, influenced the measured resistivity. Archie used the term "formation factor" (sometimes called the "formation resistivity factor") to describe this relationship, which became known as Archie's equation.

Using these relationships of resistivity and water saturation and the data from their wireline logs, engineers of the 1940s and 1950s could make fairly valid calculations of the amount of hydrocarbons present in a reservoir.

Coring

Archie's relationships were based on clean sandstones (no silt or clay mixed in with the sand). Hydrocarbon-bearing formations are not often clean sand; they are usually sands containing significant quantities of silt and clay. Limestones and dolomites also were not part of Archie's study. Therefore, something more is needed to confirm the first approximations of log-derived porosity. What could be more helpful than getting a piece of the rock itself?

The Archie in Archie's Equation

Gus Archie trained as a mining engineering and an electrical engineer but took work at an oil company (Shell), where he had the luxury of pursuing a passion. His pioneering work in the 1950s on the relationships between porosity and electrical properties is credited with the early identification of the giant Elk City Field, in Oklahoma. This episode demonstrated for the first time the role that well logging could play in identifying pay zones. Like many great thinkers, this modest and unpretentious man captured the respect of his colleagues, especially as Archie and his company shared his insights with the rest of the industry. He is credited with creating the term petrophysics, but his most famous contribution, fondly referred to by his colleagues as Archie's equation, earned him the reputation of the Father of Well Logging.

The actual porosity of a rock can be measured in the laboratory if a suitable sample can be collected from the well. This is done by *coring*. The driller replaces the drill bit with a *coring bit* that has diamond-studded teeth around the bit's perimeter and a hole in the center (fig. 6–12). As the bit is rotated, it grinds a path around a circular piece of the formation. This cylindrical piece slides up the bit into the core barrel above the bit in pieces as long as 30 feet and three to four inches in diameter. (Occasionally, when the rock is very hard and not susceptible to breaking up, two core barrels are screwed together, and cores as long as 60 feet may be recovered with only one trip into the hole.) These cores are retrieved and sent to the laboratory for firsthand porosity measurements and description of their lithology.

Cutting and retrieving a core is a time-consuming and costly operation. Moreover, it is only effective where the rock is hard enough to stay together during the coring process. In places like the U.S. Gulf of Mexico, this is not the case. In those locations, instead of using a coring bit, wireline cores

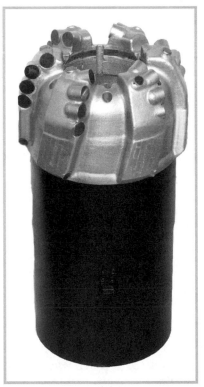

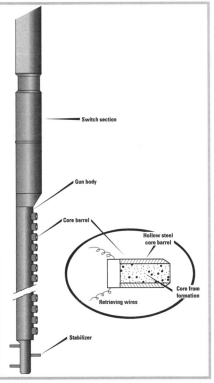

Fig. 6–12. Coring bit. The bit cuts a cylindrical column of rock that passes up the center of the bit and into a section of special pipe above the bit called a "core barrel."

Fig. 6–13. Sidewall coring gun. This device fires short, hollow tubes into the sides of the borehole capturing samples of the rock. The tubes are anchored by short wires and recovered as the entire assemblage is pulled from the hole.

are taken (fig. 6–13). This operation entails running a tool in the hole that has multiple cup-shaped receptacles along its sides that can be mashed into the sides of the hole. They are retrieved as the tool is pulled away and then upward from the sampled interval. In the best of worlds, the wireline cores come out about $3/4$ inch in diameter and as much as $1\frac{1}{2}$ inches long. More often, only partial cores are recovered, and the porosity measurement becomes more of an estimate than a true value. Sometimes firing the sampler into the formation so disturbs the material that measurement of its porosity may be only relative and probably too high.

Coring alternatives

A more routine way to determine porosity is through logging with radioactive tools. Two types of popular radioactive porosity-determining tools are the *compensated density log* and the *compensated neutron log*. Both tools use the formation's response to bombardment by radioactive material to measure either density porosity or neutron porosity. Engineers use these logs in conjunction with charts developed from laboratory studies to determine accurate values of porosity. They may also run a logging device known as a *sonic log*. This tool uses sound waves and measures the time taken for waves emitted from the sonde to pass through the formation and back to receivers in the sonde. Engineers may use a combination of the radioactive and sonic devices, depending on the sequence of rocks they are logging in a particular hole. Sequences of shale, sand, and limestone or dolomite generally need a combination of radioactive logs. When gas-filled formations are expected, the sonic log in conjunction with the radioactive logs is the logging suite of choice.

Correlation Logs

Correlation logs are run primarily to gather the data necessary to correlate subsurface features among several wells in the same vicinity. They include spontaneous potential logs and gamma ray logs.

Spontaneous potential

Early in the history of electric logging, another phenomenon was utilized—electric potential, or *spontaneous potential* (SP). When water in the mud contacts the edge of two different formations, one sand and the other shale, a small electric current is generated. A reading, again taken with a

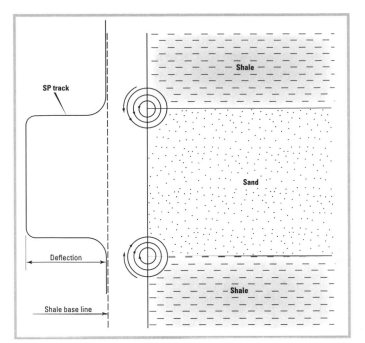

Fig. 6–14. Technology behind SP logging. The spontaneous potential or "SP" is the current developed when two different rock types contact each other in the presence of water.

sonde, delineates the top and bottom boundaries of the sand. Normally, in a sequence of sand and shale beds, the SP recording plotted against depth (called a curve) and opposite shale is a flat line (fig. 6–14). When a sand bed is traversed by the detecting tool (in a sonde), the SP curve is deflected to the left. The amount that it is deflected depends on how *clean* it is—how little shale it contains. To have SP, a permeable formation must lie next to a shale. This happens in a sequence of sand and shale, but a limestone is impermeable and highly resistive. These two characteristics prevent the development of an SP when a limestone (or dolomite) is present.

Modern logging devices have replaced the SP log in determining quantitative measurements of R_t. Today, the SP curve is used extensively to correlate logs (to identify the same formation in different wells) and to measure bed thickness.

Gamma ray

A *gamma ray log* records the natural radioactivity of shales. Gamma radiation correlates well with the SP curve and is also a good indication of the amount of shale that a sand bed may contain, as on figure 6–11. But an SP log can be recorded only in an open hole (a section without casing). A gamma ray log can be used both in the open-hole situation and after casing has been run (fig. 6–15). Correlating the two logs helps engineers determine the exact position of a target reservoir during completion operations after casing has been cemented in the hole as described later in the section on perforating.

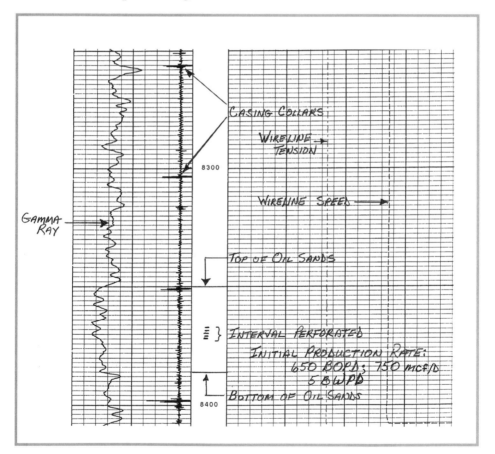

Fig. 6–15. Gamma ray. A log analyst has correlated this cased hole log with the open hole log (fig. 6-11) and determined where to perforate the casing.

Real Time

Offshore, where time is money—a lot of money—pulling the drill bit to run a sonde is always expensive. That gave incentives to develop *logging while drilling* (LWD) and *measurement while drilling* (MWD), particularly for offshore wildcat wells. Instead of a wireline sonde, logging devices are placed on the drill string, just above the drill bit. Readings from the electronic devices inside are transmitted, incredibly enough, as pulses in the mud, read by sensors at the surface, and inputted to onboard computers. The drilling team can read the results in real time. Sometimes the data are also transmitted to the operator's headquarters onshore, where other staff members monitor them and offer advice to the team on the rig. This logging technique has become standard both onshore and offshore in high-angle directional holes.

Directional and Horizontal Wells

In the early decades of drilling, wells were drilled as nearly vertical as was feasible. However, by the middle of the 20th century, technology had improved to the point that directional drilling to a target hundreds if not thousands of feet laterally away from the surface location of the drilling rig became an accepted practice. Directional drilling arrived just in time, as it became an important enabling technology for the offshore. Without it, developing whole reservoirs or other nearby accumulations would have required prohibitively costly multiple platforms.

A horizontal well is the ultimate directional well. From the surface the well is drilled nearly vertically until the target formation is approached. Then, the bit is diverted horizontally and kept within the upper and lower boundaries of the formation for hundreds and occasionally thousands of feet.

Logging in vertical and near-vertical wells—and even those approaching 50° inclination—is relatively easy. The weight of the logging sonde and the wireline itself carries the sonde to TD by the forces of gravity. But logging in a wellbore where the inclination exceeds about 50° is an entirely different matter. One common solution is to pump the sonde down the drill pipe by using quick connectors on the wireline so that as each joint of drill pipe is run in the hole an additional segment of wireline can be attached like an extension cord to the line holding the sonde. This is a slow and painstaking procedure.

An alternative becoming more popular as the technology improves is to use *LWD* (logging while drilling) logs obtained from logging devices used on the bottom of the drill string and located above the bit. If the engineers want verification that the LWD logs are accurate to the quantitative degree they require (there may be a question about data quality obtained from recording the pulsations of the drilling fluid), then a short segment of log can be run by attaching a sonde to the bottom of the drill pipe, along with a "smart box" (a computerized recording device), and logging a short portion of the hole. Normally, this log is limited to about 1,000 feet because of limitations of the memory of the downhole recorder.

Original Hydrocarbons in Place

Knowing the hydrocarbon saturation from the resistivity logs and the porosity from the suite of porosity logs, engineers can begin to approximate the amount of hydrocarbons that have been discovered. With these parameters, input from the geologists, and *volumetric formulas*, engineers calculate the amount of hydrocarbons in the formation before production begins.

For oil,

$$N = (1 - S_w) \times \text{porosity} \times A \times h \times B_o \times 7{,}758$$

and for gas,

$$G = (1 - S_w) \times \text{porosity} \times A \times h \times (1/B_g) \times 43{,}560$$

where

- N is the total oil in place (in barrels)

- G is the total gas in place (in cubic feet)

- A is the drainage area (in acres)

- h is the net pay thickness (the thickness of the formation corrected for any shale or nonporous layers)

- B_o is the oil shrinkage factor (a measure of how much oil shrinks as it reaches surface temperature and pressure [for more on this, see chapter 7])

- $1/B_g$ is the gas formation volume factor (a measure of how much gas expands when it reaches surface temperature and pressure [for more on this, see chapter 7])

- 7,758 is a constant to convert acre-feet into barrels

- 43,560 is a constant to convert acres into square feet

For example, suppose an oil-bearing formation has been discovered. The parameters required in order to calculate the amount of oil discovered and their sources are as follows:

- Water saturation, S_w, is 15% determined by examination of the resistivity from induction logs.

- Porosity is 25% (0.25) determined by examination of cores and/or a combination of density, neutron, or sonic logs.

- The drainage area (the productive area), obtained from the geologist's maps, equals 250 acres.

- The net pay thickness found on the gamma ray or SP logs is 27 feet.

- The oil shrinkage factor is assumed, on the basis of experience and laboratory work with oils common to the area, to be 0.6.

The solution becomes

$$N = (1 - 0.15) \times 0.25 \times 250 \times 27 \times 0.6 \times 7{,}758$$

$$= 0.85 \times 7{,}894{,}975$$

$$= 6{,}676{,}729 \text{ barrels}$$

In a similar fashion, the volume of gas discovered can be calculated.

The calculation gives an estimate of the total amount of hydrocarbons in the reservoir. But that is not the amount the engineers expect to be able to produce. That requires another set of assumptions, which result in a *recovery factor*. In simple terms, the recovery factor is the amount of oil to be recovered (produced) as compared to (divided by) the total oil in the reservoir (*N* in the above equation). In most cases, the volume produced (the ultimate recovery) is less than 50% of the total hydrocarbons in the reservoir; often it is on the order of 30%, especially in oil reservoirs.

Open-Hole Testing

Drillstem testing

Exploration managers and investors are always excited by evidence of a successful well—a discovery—as told by logs and shows. However, nothing is sure until oil or gas is actually produced from the well. When analysis of the mud indicates a hydrocarbon show of sufficient strength and the penetration rate of the bit suggests penetration of a sand that may be a reservoir rock, geologists often call for a *drillstem test* (DST). This exercise tests an interval, segregated from the rest of the hole, that appears to be hydrocarbon bearing.

The simplest, most straightforward test is when the interval to be tested is at the bottom of the hole. Normally, the driller, with help from the geologist, tries to drill only a few feet of the target horizon in order to minimize the chance of drilling through the oil-bearing portion of the reservoir into what might hold water.

For the test, the driller pulls the drill string out of the hole and replaces the bit with a special test assembly, *or tool*, which is then run back into the hole on the end of the drill pipe. In its simplest form, the DST assembly consists of a valve, a packer, and a pressure recorder, with a section of perforated pipe at the bottom, through which formation fluids can flow (fig. 6–16).

The driller runs the drill string to the bottom of the hole with the valve closed. This prevents any drilling mud or other fluid from entering the drill pipe. The pressure recorder, located below the closed valve, records the increasing weight of the column of mud above it (the hydrostatic pressure), on the outside of the drill string, as it goes down the hole. When the DST assembly reaches total depth (TD), the driller lets a small amount of weight rest on the tool. This weight compresses the DST assembly and causes the packer elements to expand, sealing off the hole and separating what is above the packer from the interval to be tested. With the valve closed

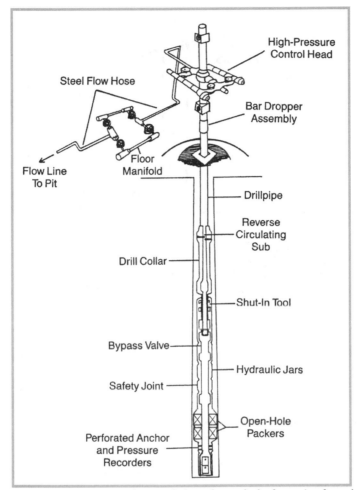

Fig. 6–16. Tools used in DST. The open hole packer seals the formation from the mud column and allows formation fluids to flow through the drill pipe to the surface.

and the packer set, the recorder begins to detect formation pressure, which at this point is abnormally high owing to compression of the drilling fluid in the wellbore caused by expansion of the packer. This condition is referred to as "supercharged" and is corrected by opening the test valve briefly, to release the trapped pressure. This is referred to as the "preflow period," usually lasting only two or three minutes. The test valve is closed again, and the recorder begins to detect formation pressure, which hopefully builds toward a reading of its actual pressure (fig. 6–17). If the permeability is high, the formation pressure is reached very quickly; if the permeability is low, a true reading of the formation pressure may not be obtained. This

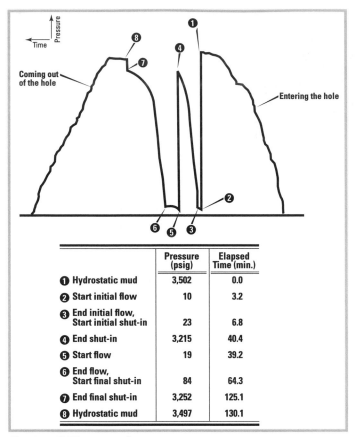

	Pressure (psig)	Elapsed Time (min.)
❶ Hydrostatic mud	3,502	0.0
❷ Start initial flow	10	3.2
❸ End initial flow, Start initial shut-in	23	6.8
❹ End shut-in	3,215	40.4
❺ Start flow	19	39.2
❻ End flow, Start final shut-in	84	64.3
❼ End final shut-in	3,252	125.1
❽ Hydrostatic mud	3,497	130.1

Fig. 6–17. DST pressure chart

part of the test lasts about a half hour. The recorded pressure is called the *initial shut-in pressure.*

The driller now rotates the drill string slowly, and the valve opens again, allowing the fluids in the formation to flow into the drill pipe. Excited anticipation spreads over the rig floor at this defining moment of the wildcat well.

The pipe at the top of the drill string is partially open. The rig personnel can hear and feel the air in the drill pipe being pushed out by the incoming fluid. If there are hydrocarbons in the reservoir, they arrive with a rush, provided that the permeability of the formation is enough to let them in quickly. With a good, high-permeability formation, the hydrocarbons may flow to the surface within the 30–60 minutes the driller leaves the valve open. Keeping it open any longer is risky because the entire drill string may become stuck. Drillers become nervous whenever pipe is in the hole and not moving—and especially when there is no circulating mud flow. The longer the packers on the DST tool remain expanded and pushing against the sides of the hole, the less likely it becomes that they can be released and pulled free, a lesson learned from hard experience.

If and when hydrocarbons arrive at the surface during a test, pollution concerns dictate their handling. At an onshore operation, oil is directed to a storage pit specially dug for the purpose. Gas is sent down a pipe, away from the rig, and ignited, an operation called *flaring*.

Offshore, when gas flows to the surface during a test, it is also diverted by a long pipe away from the rig and completely burned, just as on land. Testing a potentially oil-bearing horizon is approached much more carefully. The flow is directed to an onboard tank that has limited capacity, so the test period is confined to just long enough to determine the type of hydrocarbon.

Following the flowing phase of the DST, the driller closes the valve. Once again the pressure recorder senses and records the formation pressure (the *final shut in pressure*). That may or may not be the same as the initial value. The driller unseats the packer and lifts the drill string off bottom. His weight indicators tell how much fluid is contained in the pipe. During the test, hydrocarbons may not have flowed all the way to the surface. The formation may be too *tight* (too low a permeability) to give up enough fluid. Maybe only water or both oil and water are in the pipe. Cash usually passes between hands of the rig crew to settle wagers as the drill pipe is pulled from the hole and the top of the fluid is found.

Besides the flow of hydrocarbons, of equally great interest to geologists and their management are the pressure-recorder readings. They show how fast the pressure built up during the initial shut-in time and what pressure was recorded then. These readings are compared to the final pressure recorded after the flow period ended. These data provide the basis for calculation of reservoir pressure, an indication of whether any damage from mud filtrate has occurred (see the section on drilling mud in chapter 5), and some important clues as to the size of the discovery.

Suppose the well flowed 40 barrels of oil and the difference between the initial pressure and the final pressure is, say a difference of 20%. The engineer concludes that the reservoir is in reality a very small compartment. Because of the pressure decline, 40 barrels represents 20% of the total recoverable hydrocarbons. The total left in the reservoir would be about

160 barrels—a very disappointing discovery! Fortuitously, in the majority of DSTs, the pressures in the reservoir both before and after the test are so nearly the same that no depletion of the reservoir can be discerned.

Wireline drillstem tests

Since cost, chance of pollution, fear of sticking the pipe, and hole deviation all present problems in testing, DST tools are often run in the hole by using a wireline, rather than drill pipe (fig. 6–18). This operation uses a tool with interior chambers and a pressure recorder run in the hole on a cable. The bundle of wires in the cable performs three functions:

- suspends the tool

- activates it

- transmits recorded data to the surface for processing

Fig. 6–18. Portion of a wireline formation testing tool. The port and its surrounding packer are shown in an extended position. A sample chamber to hold produced fluid is screwed to the bottom of the tool. Courtesy of Baker Atlas.

On one side of the tool is a spring-loaded pad. When extended, it pushes the tool against the opposite side of the hole, against the interval to be tested. A signal down one of the wirelines opens a small port. Fluid, oil, gas, water, or a combination enters one of the chambers in the test tool. Simultaneously, the pressure in the test interval is recorded. The routine is usually repeated at different levels. When the tool is returned to the surface, the samples are recovered in pressure vessels for analysis.

Petroleum engineers use the DST pressure and fluid-recovery data, information from the logs, and volumetric formulas in order to calculate the amount of oil and gas in the formation. From that calculation, they can determine if running casing and completing the well will be economic.

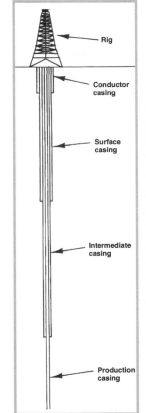

Fig. 6–19. Types of casing. As the well is deepened, successive strings of casing are run each inside the previous string. Each drilling bit must pass through the smallest diameter casing. As the hole gets deeper, the borehole becomes smaller.

Completions

If the operator decides there is enough oil and or gas present to justify the cost of producing the discovery, the next step is to case the hole. Without casing and without mud circulating, the open hole will collapse. In addition, without casing, all the water-bearing intervals penetrated above the discovery interval and not already cased off would contribute to production. Some hydrocarbons could even invade and pollute aquifers at higher levels. Therefore, the last part of the wellbore has to be cased. *Production casing* is run and cemented inside whatever strings of casing were used to drill to TD (fig. 6–19). Normally, production casing is between 4½ and 7 inches in diameter. Later, *production tubing* will be run inside that casing.

Solid or slotted

Before the production casing is run (and often even before the final part of the hole is drilled), a decision is made about whether this final string is to be a solid string of steel pipe or to be combined with a *liner* (a slotted and perhaps wire-wrapped pipe) positioned across the interval to be produced. The decision depends in large part on whether the producing formation is very loosely consolidated sand that will flow into the perforations and plug up the tubing string. In this case, the slotted liner will be used, and a *gravel pack* operation (described below) will be performed as part of the completion. If, by contrast, the formation is solid and will stay in place throughout the life of the well, a solid casing string is run, cemented, and perforated across the producing zone.

Cementing

With a solid string of production casing in place, a cement slurry is pumped down the inside of the casing and up the annular space between the back side of the casing and the hole and is lapped over the

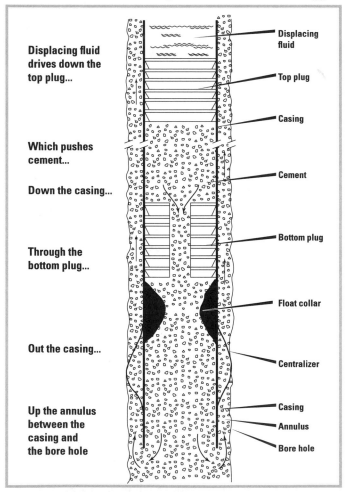

Displacing fluid drives down the top plug...

Which pushes cement...

Down the casing...

Through the bottom plug...

Out the casing...

Up the annulus between the casing and the bore hole

Displacing fluid

Top plug

Casing

Cement

Bottom plug

Float collar

Centralizer

Casing

Annulus

Bore hole

Fig. 6–20. Tools used in a cementing operation. Cement is pumped down the drill pipe followed by wiper plugs, followed by water or mud. The plugs are caught at the bottom of the drill pipe leaving cement on the outside of the pipe and water (or mud) on the inside.

bottom of the previous casing (fig. 6–20). If properly mixed and pumped, the hardened cement isolates all water or hydrocarbon-producing horizons from each other. The casing and its surrounding cement sheath must then be perforated opposite the hydrocarbon intervals so that the fluids in the reservoir can flow into the wellbore. The next operation is to provide the surface termination of the production casing string and the devices through which the production tubing is run into the wellbore.

Trees

After the production casing (the last casing string) has been cemented, a *wellhead* is installed. This combination of steel flanges, spools, and hangers bolted together is designed to provide surface control of the subsurface fluids. The production tubing is run through the wellhead and secured by a tubing hanger.

The surface controlling valves are bolted to the top of the wellhead. Onshore, their design ranges from simple to complex, depending upon the pressure of the producing formation, whether it produces oil or gas, and the number of tubing strings run in the wellbore (single or multiple completion).

If the combination of surface controls is more elaborate, the assembly is often called a *Christmas tree* (fig. 6–21). Stories abound about the use of the word *Christmas*, but the appearance of the valves,

Fig. 6–21. Christmas tree for a high-pressure gas well. Wheels are used to open and close the valves which control production rates. Courtesy of Cooper Cameron.

handles, and gauges allows an easy mental leap to the term *tree*. In its simplest form, the surface device is known simply as a wellhead or tree.

Wells drilled offshore can be connected to either of two types of trees:

- dry trees
- wet trees

Dry trees sit on a fixed platform that is attached to the seabed. They look and are operated like onshore trees, providing the same production control mechanisms.

Wet trees sit on the seabed and are attached to the top of the well, just as dry trees are. However, because of their subsea environment, the valves and control mechanisms are encased in a solid steel box and connected to the surface production facility by both hydraulic and electric systems for activation (fig. 6–22). The enclosing steel protects the valves and control systems from the seawater and subsea pressures. Also provided are quick-connect fittings that allow remote operated vehicles to connect to and manipulate the valves and controls. That can be necessitated by failure of the electric and hydraulic connections between the tree and the host platform or when maintenance is required.

From the wet tree, the production flow line can extend to a subsea manifold or a subsea pipeline or a riser connecting it to a platform.

Wet trees are more likely to be used in deeper waters when the wells need to be connected to floating platforms: tension-leg platforms; floating production ships; floating production, storage, and off-loading vessels; and spars. (More about these *floaters* is provided in chapter 9.) Wet trees are sometimes used when a field is too small to warrant installing a fixed production platform. In that case, they can be connected by subsea pipeline to a fixed platform that might be 60 or more miles away. Once the surface control assembly, the tree, has been installed, the next step is to prepare the well to produce.

Perforating

If the production casing has been run and cemented across the hydrocarbon-bearing interval, it needs to have holes punched in it and its surrounding cement sheath in order to permit the hydrocarbons to flow into the wellbore. It needs to be perforated.

The device for perforating the casing is known as a perforating gun. It is usually run on a wireline and in conjunction with a gamma ray/casing-

Fig. 6–22. Wet tree. This assembly of well valves is the offshore equivalent of the tree pictured in figure 6–21. It is set on the ocean floor and controlled by hydraulic and electric lines to the surface. Courtesy of Drilquip.

collar locator assembly. The gun punctures the casing and the cement sheath surrounding the casing either by firing bullets from a set of small guns spaced about six inches apart or by launching shaped charges called jets (fig. 6–23). (Jet charge perforating grew out of the discovery, during World War II, that shaped charges of explosive powder, ignited against the armor plating of tanks, produced a far more effective hole than did a solid steel projectile.)

The gun and its locating apparatus are lowered through the production tubing, which has been set with a packer just above the pay interval, the perforation target. Most of the fluid in the tubing is bailed from the hole or is circulated out with foam and coiled tubing (described in chapter 10). Only enough fluid is left to dampen the upward surge of the perforating gun as the formation fluids rush into the casing and up the tubing.

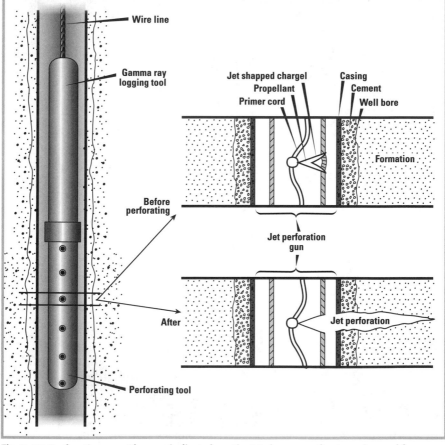

Fig. 6–23. Perforating gun. Blowups indicate how the jet charges perforate casing and form flow channels into the formation.

The intervals to be perforated have been previously defined by the open-hole logs, and the intent is to perforate the casing precisely opposite those intervals. But most of those logs cannot see through the casing that has just been placed in the hole. To locate the perforation target, a gamma ray–collar correlation log is usually used (fig. 6–15). This includes a gamma ray log plus a record of the location of casing collars, the points at which two joints of casing are screwed together. The gamma ray log can be correlated with the log run in the open hole. With this information, the engineer can pinpoint where the hydrocarbon horizons are by correlation with the open-hole logs. Directions to these spots are given in terms of the location of casing collars, which are themselves identified by distances from particular depths in the well.

Suppose that the open-hole log showed the hydrocarbon interval to be at a depth of 8,351–8,384 feet. The cased-hole gamma ray/casing-collar log showed the same interval between 8,350 feet and 8,386 feet. The difference between the depths shown by the two logs is due to the use of different wirelines and different logging devices. But the difference is not important because the engineers recognize the signal on the gamma ray log of a casing collar at 8,304 feet and another signal at 8,351 feet. They can describe the target interval as being 15 feet below the collar at 8,351 feet and extending 6 feet below that point.

When the gun reaches the correct position, the signal is sent through the wireline to fire the charges, perforating the casing and allowing reservoir fluid to flow in and up the production tubing.

Gravel Packing

All of the above comprises the completion technique of choice where the formations are solid enough to stay in place. But in much of the offshore Gulf of Mexico; in many locales in California, Venezuela, Nigeria; and, in fact, in altogether too many places in the world, completion matters are complicated (and made far more costly) by having to prevent sand from invading the producing system. The system is known as *gravel packing*. It is mechanically more complicated than simply cementing the production casing in the hole.

In simple terms, a slurry of properly sized coarse sand (the gravel) is circulated on the outside of a slotted liner between it and the hole (fig. 6–24). The slots in the liner (and the wire screens if used) are of such a width that the sand grains from the gravel pack will be kept outside the liner. The size of the gravel pack particles will be such that any fine to coarse material

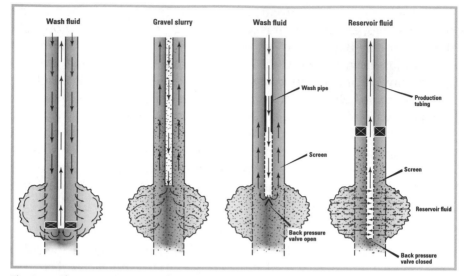

Fig. 6–24. The successive stages of a gravel pack operation

from the formation will remain within the pack. Amazingly, this gravel pack operation is performed not only in vertical and near-vertical wells but also in horizontal wells both onshore and offshore.

Stimulation

Now that the formation fluids have been provided access into the well through either perforations or a gravel-packed liner, it is time to be certain that the well will produce at its maximum potential. During drilling and completion, the formation face (exposed surface) has been subjected to all sorts of fluids, drilling mud, and various circulating media. The drilling mud will have left a thin sheath of filter cake on the formation, and mud filtrate may have penetrated as much as several feet back into the pay intervals. This filter cake and filtrate form a *skin* that severely inhibits production—the permeability of the formation could be drastically impaired. At this point, the production team may decide to *stimulate* the well, a procedure described in chapter 10 (on workovers), or to produce it to better judge the degree that stimulation may be needed and to gather data for designing the operation.

Cased-Hole Logging and Measuring Devices

In addition to the logging tools already described, a variety of special-purpose cased-hole logging tools are available:

- cement bond log—determines the quality of the cement job
- casing-collar locator log—provides depth control
- temperature log—indicates zones producing in contrast to those not producing
- casing inspection log—determines amounts of corrosion by changes in casing diameter
- flowmeter—detects changes in production inflow rates at various intervals
- radioactive tracer—determines fluid movement
- acoustic image log—determines pits and corrosion on both the outer diameter (OD) and the inner diameter (ID) of the casing
- through-casing evaluation tools:
  - gamma ray spectrometry
  - compensated neutron log
  - acoustic log
  - density log
  - pulsed neutron spectrometry log
  - pulsed neutron decay-rate log

And On . . .

Logging, DSTs, setting production casing, and perforating are not the only steps engineers need to take to complete a well. Testing, stimulation, and water-inflow control follow in a few chapters. But to deal with those, engineers need to know a little about the conditions in the reservoir and how the properties of the hydrocarbons will affect their behavior during production.

7

Behavior: Hydrocarbon Activity in the Reservoir

We will now discuss in a little more detail the struggle for existence.

—Charles Darwin (1809–1882), *The Origin of Species*

When bartenders open a magnum of champagne, especially after jiggling it, they get one effect. If they open a bottle of whiskey, they get another, less messy effect. And if some festive partygoers blow up some balloons and then let them go, they get a third effect, as they bend over in fits of glee. Tapping oil and gas reservoirs has some similar, more serious characteristics.

The producing phase of oil and gas operations is a chapter away, but an important prerequisite is some understanding of what happens to the hydrocarbons in a reservoir during production. The removal of some of the contained fluids affects the amount of recovery of the rest and calls for tailored production techniques.

Phases

As far as petroleum engineers are concerned, when a hydrocarbon accumulation is discovered, it exists in a virgin condition. It may have not only liquids of a unique composition but also gases dissolved in the oil and/or in a free state.

Every hydrocarbon accumulation has a unique history—the makeup of its container, the organisms of its origins, the "fermentation," the migration, and the temperature and pressure changes from tectonic movements, both vertical and horizontal. It is no wonder each has its own composition and fingerprint.

The behavior of the fluids in the reservoir depends on this unique composition. Petroleum engineers have a tool, the *phase diagram*, that is a useful way to present how the hydrocarbons in a reservoir will respond as production proceeds. As an example of phases, take the substance water (H_2O). It has three phases, liquid, crystalline, and gas—water, ice, and vapor or steam. A glass of ice water is the classic three-phase, coexistent system. (There is always a little vapor above the water.)

Phase Diagrams

Consider the phase diagram in figure 7–1 for a two-phase hydrocarbon system, liquid and gas. The chart shows the possible states in the reservoir, including a prominent area called a *saturation envelope*. A *bubble point line* bounds the envelope on the left; a *dew point line* forms the right boundary.

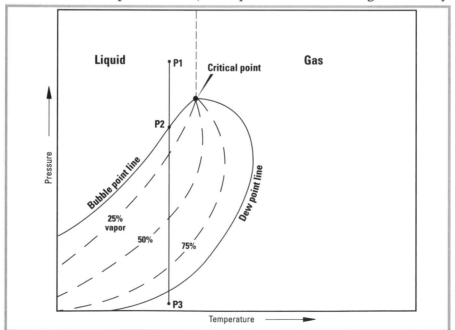

Fig. 7–1. Typical oil and gas phase diagram. The iso-thermal line from P1 to P3 passes from an all-liquid phase to liquid-vapor phases to an all-vapor phase.

The Funny Thing about Oil and Gas

For the record, because oil and gas are mixtures of numerous (sometimes hundreds of thousands) different chemicals, their phase diagrams differ from the phase diagrams of some pure substances like water and carbon dioxide.

Pure substances like these do not have broad areas of coexistent phases like hydrocarbons. They typically change from, say, vapor to liquid at a defined pressure-temperature boundary, without saturation envelopes or iso-mixture lines. And above the critical temperature, the fluid becomes *supercritical*: no matter how high the pressure, the vapor will not condense to liquid.

They meet at the *critical point*. Inside the envelope, various mixtures of liquid and gas exist, indicated by the *iso-mixture* lines and dependent on the pressures and temperatures. Outside the envelope, to the left of the critical point, only liquid exists; to the right, only gas exists.

What happens in the reservoir when the pressure is reduced while the temperature remains constant? (Normally, there is no change in temperature in a reservoir as the pressure is lowered.)

Follow the *isotherm*, the constant temperature, line in figure 7–1, as the pressure in the reservoir drops from point P1 (in the liquid phase) to point P2. As it passes P2, the bubble point, the very first bubbles of gas form. Follow the line through the saturation envelope as it crosses the iso-mixture lines of percent gas and liquid to point P3, the dew point. There, the last drop of liquid vaporizes. Outside the dew point line, nothing but gas exists in this system.

If the initial reservoir pressure (the virgin pressure) is above the bubble point, the oil (the liquid) is said to be *undersaturated*—it could hold more gas. The long history of this reservoir resulted in more large, heavy hydrocarbon molecules (liquid oil) than smaller, lighter molecules (gas).

If the initial reservoir pressure is below the bubble point pressure, the liquid oil is *saturated*—it can hold no more gas. Any free gas forms a *gas cap* (an accumulation of gas at the top of the reservoir) because the gas, being less dense than the oil, has floated to the top by the forces of gravity.

Reservoir Fluid Categories

Petroleum engineers recognize five major types of reservoir fluids:

- black oil

- volatile oil

- retrograde gas

- wet gas

- dry gas

This engineering nomenclature is not to be confused with the chemical and commercial descriptions of hydrocarbons in chapter 3 or any other contradictory names that regulatory agencies might use. These names are used because they relate to hydrocarbons as they exist in the reservoir.

Each of these fluid types responds in its own way to changes in temperature and pressure. Each can be characterized by its own phase diagram.

Petroleum engineers need to consider the specific phase diagram before earnest production begins. It is the deciding factor in many of the decisions concerning production management. Knowing it affects selection of surface equipment, oil-gas mixture calculations, prediction of oil and gas reserves, depletion planning, well spacing, and the choice of the appropriate enhanced recovery technique.

After a newly discovered reservoir has been found, an early and important step is to obtain a sample of the hydrocarbons, known as a *PVT sample* (pressure, volume, temperature). This is collected in a pressurized container during a drillstem test or at first production and delivered to a laboratory. There, the gas and oil can be recombined at conditions as close to virgin as possible. From the laboratory analysis, the phase diagram *for the oil in that reservoir* can be constructed. Also from the samples, the type of hydrocarbon in the reservoir can be reliably identified by observing three critical characteristics of the produced fluids in table 7–1:

- the initial producing *gas-oil ratio* (GOR)

- the *gravity* of the liquid

- the *color* of the liquid

Table 7–1. Field characteristics of fluid types

Fluid Type	Gas-Oil Ratio	Gravity	Liquid Color
Black Oil	<2000 scf/STB	<45 degrees API	Very dark, often black, but sometimes green or brown
Volatile Oils	2000 to 3300 scf / STB	>40 degrees API	Brown, orange, sometimes green
Retrograde Gas	3300 to 50,000 scf / STB	40 to 60 degrees API	Light colors, brown, orange, green to clear white
Wet Gas	>50,000 scf / STB	40 to 60 degrees API	Clear white
Dry Gas	No liquid hydrocarbons formed		

Using Phase Diagrams

Petroleum engineers use phase diagrams to anticipate what will happen to the hydrocarbons in the reservoir as they produce them and inevitably lower the reservoir pressure. In the following characterizations, the initial reservoir pressure is denoted P1 and is above the saturation envelope. It could just as well be at a lower pressure, even lower than P2, but the conclusions would be the same.

Black oil

The reservoir pressure in the black oil phase diagram (fig. 7–2) at P1 is above the bubble point pressure at P2 and to the left of the critical point. Only liquid oil resides in the reservoir at this point. The oil is undersaturated (more gas could have been dissolved in it).

As the pressure in the reservoir falls owing to production (along the line P2 toward P3 in fig. 7–2), free gas evolves in the reservoir—it starts to separate, just as carbonation bubbles seem to appear from nowhere in a glass of champagne. The percent liquid decreases as the isothermal line passes from one *iso-mixture* line to another.

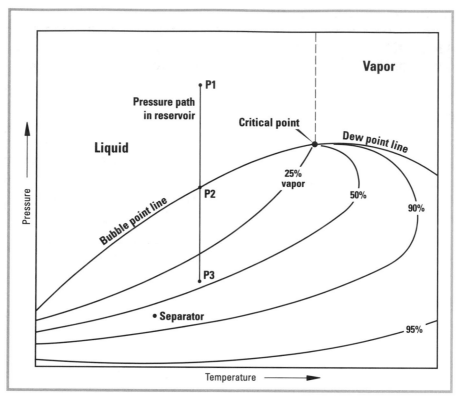

Fig. 7–2. Typical black oil phase diagram

The pressure/temperature of the surface separator is also shown in figure 7–2. Since it lies within the phase envelope, a large amount of liquid hydrocarbons are produced into the separator. At initial production, the stream coming out of the well is all liquid oil, but as the hydrocarbons pass up the tubing and into the (lower-pressure) separator, the gas vaporizes out of the oil. As the reservoir is produced and the reservoir pressure falls below the bubble point, the amount of gas going into the separator continues to increase. Consequently, the GOR of the production stream available for sale increases correspondingly.

Volatile oil

The saturation envelope in the phase diagram for volatile oil (fig. 7–3) has a quite different shape. The critical point of the volatile oil graph is closer to the reservoir temperature—it has shifted to the left. Also, the iso-mixture lines are not as evenly spaced. They are shifted closer to the bubble point line. This is an indication that gas will evolve much more

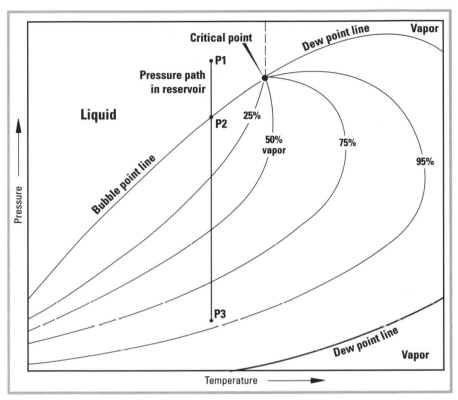

Fig. 7–3. Typical volatile oil phase diagram

rapidly with a decline in pressure from P2 to P3 in a reservoir containing volatile oil than for a black oil reservoir.

More gas is produced sooner in the life of this reservoir. However, the stream is still rich in liquids. As much as 50% of the produced liquids enter the separator as gas, to be separated there.

Retrograde gas

A discovery that proves to contain retrograde gas is usually a financial disappointment. A great deal of the liquids that condense from the gas in the reservoir may be unproducible, and the cost of maintaining high reservoir pressure in order to keep them in the vapor phase is high. Fortunately, retrograde gas reservoirs are fairly uncommon.

Comparing figure 7–4 with the prior two phase diagrams shows that in this reservoir, the critical point temperature is less than the reservoir temperature. At virgin conditions, all the hydrocarbons are in a gaseous state.

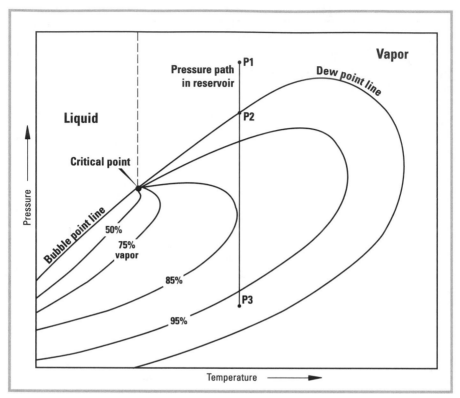

Fig. 7–4. Typical retrograde gas phase diagram

As the pressure falls in the reservoir owing to production (from point P1 through the dew point P2 in fig. 7–4), liquid starts to condense in the reservoir, unfortunately most of it away from the borehole. Because of the *relative permeability* (a subject to be covered later), the gas flows through the pores, but most of the liquid does not. Much of it adheres to the walls of the pore spaces. The liquid left behind could make up as much as 35% of the total oil in place and cannot be recovered under normal circumstances. One solution is to recycle gas from the separator back to the reservoir to maintain the pressure as high as possible, keeping more oil in the gaseous state.

Wet gas

The "wet" in wet gas connotes a modest concentration (10–25%) of hydrocarbons heavier than methane, but in a gaseous state. That does not include water, since all reservoir gas is saturated with water, another topic to be dealt with later.

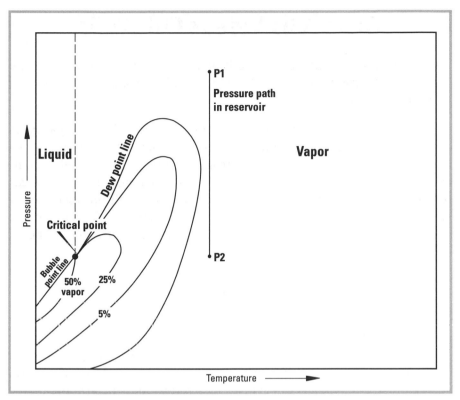

Fig. 7–5. Typical wet gas phase diagram

The saturation envelope (fig. 7–5) for a wet gas is smaller than that for any previous fluid type. It is so compact that the entire envelope is to the left of the isothermal line described by the fall in pressure from original conditions, from P1 to P2. For the entire producing history of the reservoir, nothing but gas exists. However, conditions at the separator are such that some liquids drop out of the gas stream there. These are the colorless liquids referred to commercially as condensate.

Dry gas

The saturation envelope for dry gas is even tighter than figure 7–5 and farther from the pressure decline line from P1 to P2. The critical point is at a lower temperature and pressure than the temperature or P2 pressure in the reservoir. Consequently, as the pressure in the reservoir declines, no liquids evolve in either the formation or the separator. The produced gas is almost entirely methane.

Gas Wells Versus Oil Wells

The first two categories, black oils and volatile oils, produce through wells generally called oil wells, although volatile oils can result in a large proportion of *associated gas.*

The wells tapping the reservoirs containing any of the three gas types described earlier can be and often are called gas wells. All three reservoirs hold gas at initial conditions. The retrograde reservoir ultimately retains most of its liquids, producing a relatively small amount. The wet gas reservoir produces condensate, and the dry gas reservoir produces nearly 100% gas. All the wells produce dissolved or entrained water to the surface separators—some dissolved in the gas, some entrained in the oil and gas streams.

Shrinkage

As hydrocarbons move from the reservoir to the surface, they experience declining pressure; as a consequence, they change in volume. The release of pressure might be expected to result in expansion. Not so. As the pressure declines, gas evolves (separates) and the volume of the liquid contracts (shrinks).

All liquid hydrocarbons in reservoirs contain some dissolved gas, so all liquids shrink to some extent. The oil's *formation volume factor* (FVF) compares a unit volume of a reservoir barrel at initial conditions to that of oil produced at the surface (a *stock tank barrel*). FVFs always have a value greater than 1.0. A typical black oil has an FVF of 1.47. Thought of in another way, the reciprocal of this number (1/1.47) is the oil's *shrinkage factor* (in this case, 0.68). A stock tank barrel of this oil is only 68% of the volume occupied in the reservoir at virgin conditions. The lost volume has emerged from the separator as gas.

By contrast, dry gas does expand, and as might be expected, its relationship to reservoir unit volume is called a *gas expansion factor*. With a few volumetric constants thrown in, it is the reciprocal of the liquid's FVF and equivalent in theory to the shrinkage factor.

Solution Gas-Oil Ratio

Knowing the volume of oil in the reservoir, its GOR, and the information from the phase diagram provides petroleum engineers the essentials to calculate the amount of gas that will be liberated during production of the oil phase. Solution GOR is the amount of gas produced from the separator at the surface as compared to the amount of oil produced. Both are measured at the operating pressure and temperature of the separator and converted to standard conditions. The units are standard cubic feet per stock tank barrels (SCF/STB). Reasonable values for GOR of black oils are between 475 and 1,600 SCF/STB.

Relative Permeability

There is another, particularly arcane concept pertaining to fluid movement in the reservoir that should be appreciated but need not necessarily be understood in strict scientific or mathematical terms: relative permeability. When Darcy made his calculations of reservoir permeability (described in chapter 2) for the flow of water in a sand reservoir, the reservoir was 100% water saturated. There was no other fluid present. This measure of permeability is called *absolute* permeability, and in this case, it was absolute water permeability.

When the pore space of a rock contains both oil and water, as it often does in a reservoir, the oil has one permeability value, the water has another, and both depend on the relative amounts present (their relative saturations). When water saturations are high, the water will move through the reservoir faster than the oil. The opposite is true when oil saturations are high.

Recall from chapter 2 that oil migrates upward from the source rock toward the surface through porous rock filled with the saltwater in which the sediments were deposited. It displaces water as it moves up. When the oil is finally trapped, it continues to displace water until no further displacement can proceed. At this point, unfortunately, not all of the water is forced out of the pore space. It is held in place by surface tension on the rock grains and by capillary forces. This irreducible saturation value is called the *connate water saturation*, and it may approach 20% in many oil reservoirs. It depends on the size of the pores and the sorting (clean vs. poorly sorted) of the reservoir rock.

By definition, relative permeability is the *relationship of the permeability of a fluid in the presence of another fluid as compared to the first fluid's permeability without any other fluid present, expressed as a decimal.*

On figure 7–6 are plotted typical relative-permeability curves for a particular rock. Dropping from 100% saturation on the water curve on the right down the curve to 85% (a 15% increase in oil saturation) sharply reduces the relative permeability of this rock to water, from 100% to 60%. At the same time, even though the oil saturation has increased to 15%, the rock's relative oil permeability is still essentially zero. Oil is present in the pore space, but it does not move under normal circumstances.

The 15% oil saturation in this example is a threshold called the *critical saturation*. Above that percent oil saturation, oil begins to move. From an equally important perspective, it is also called the *residual oil*

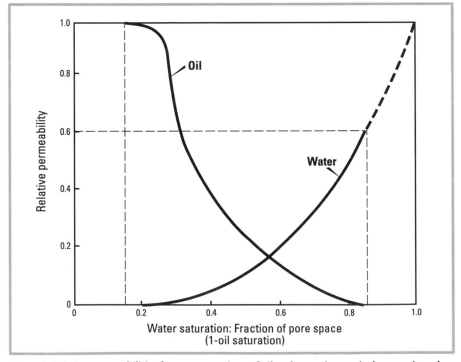

Fig. 7–6. Relative permeabilities for concentrations of oil and water in a typical reservoir rock

saturation—below that saturation level, oil does not move. For a reservoir to be producible, enough oil has to have migrated into the formation to get above the critical saturation. As the formation is depleted by production and as the oil saturation approaches residual oil saturation, oil production stops, and the party is over. Residual oil saturation is similar to connate water saturation in that neither oil nor water can be reduced below these values in an oil-water system.

Gas-oil and gas-water systems behave similarly to the oil-water system described above.

What does all this mean with regard to oil and gas production?

- It explains why allowing water to build up in the wellbore of a gas well—which can happen for any number of reasons—is a poor practice. If after a while, the formation imbibes the water, then the formation's relative permeability will change and reduce or even stop the flow of the gas.

- It explains why, when either water or gas is drawn into an oil producing interval (coning, as described in chap. 10), the oil production rate may be permanently impaired. Likewise, the gas production rate is adversely affected when water is drawn into a gas well.

- It explains why such a significant portion of the oil (often well above 40%) in a reservoir cannot be swept out and recovered, even with waterflooding. Some exotic, non–water-oil system may have to be applied, such as those described in chapter 8, under "Enhanced Oil Recovery."

- Finally, relative permeability is the phenomenon behind the poor performance of retrograde gas reservoirs. In this situation, the heavier gas liquids condense in the reservoir, increasing the oil saturation from nothing through the residual saturation value and higher. As a consequence, the gas saturation declines, and its relative permeability falls. The gas then moves more slowly through the reservoir, and as the reservoir pressure is depleted, large amounts of both gas and oil are left behind.

Now, on to production.

Here It Comes: Production

The vapours weep their burthen to the ground.

—**Alfred, Lord Tennyson (1809–1892),** *Tithonius*

The discovery well has been cased and perforated, and production tubing has been run. The well is ready to produce and generate money—sometimes more, sometimes less. Production can come in several stages, or *phases*, depending on the character of the reservoir and the fluids that reside in it:

- the (relatively) flush flow of primary production that includes both natural flow and artificial lift

- the boost from secondary recovery

- the extra, later volume from enhanced or tertiary recovery

The Motivating Force

It is not all about pressure, but pressure is the primary motivating force. Oil and gas and associated water come to the surface because the pressure in the reservoir is greater than that exerted by the column of fluid standing in the production tubing between the perforations and the surface. After all, a column of 30°API crude from a well drilled to 10,000 feet puts a pressure of 3,792 pounds per square inch (psi) on the bottom of the well. The pressure in the reservoir is at least equivalent to the hydrostatic pressure, or 4,500 psi.

Typically, the flow is helped by the evolution of gas in the fluid column. As fluid travels up the tubing, gas in the fluid expands, lowering the density of the fluid at that point and at the same time decreasing the pressure exerted below it, all the way down to the perforations (fig. 8–1).

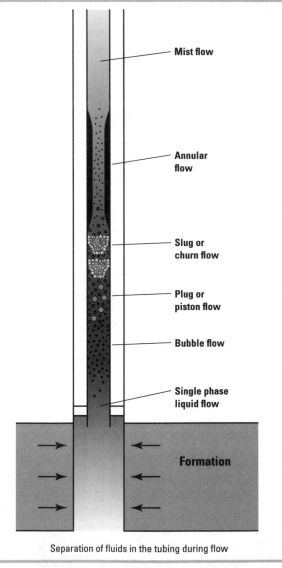

Separation of fluids in the tubing during flow

Fig. 8–1. Gas evolution in the tubing. As fluid rises in the tubing, dissolved gas expands and lowers the fluid's density (lightens it) helping it move to the surface.

As long as the pressure in the reservoir is above the bubble point, gas evolves only in the tubing or the surface facilities, but not in the reservoir. As production proceeds, the declining pressure in the reservoir will reach the bubble point, allowing gas to evolve in the reservoir, as well as in the tubing.

Thinking of pressure as energy helps in the realization of how important it is to conserve and replace it in order to recover the maximum amount of hydrocarbons from the reservoir. From the moment of first production, pressure (reservoir energy) begins to decline. Below the surface, pressure is lowest at the wellbore and highest at the farthest reaches of the reservoir (fig. 8–2). The pressure draw-down at the

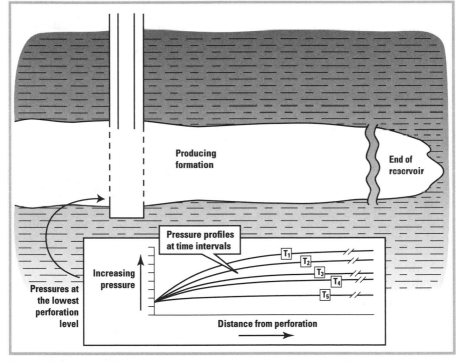

Fig. 8–2. Pressure profile in the reservoir over time. Pressure near the perforations is lower than it is a distance back in the formation. As production progresses, the entire reservoir pressure is lowered (lines T_1 to T_5). But as long as the well is producing, the pressure at the perforations is always the lowest pressure in the reservoir.

wellbore progresses as a wave to the farthest point. If nothing replaces the fluids as the reservoir is depleted, the pressure and the production rate decline. The decline rate depends in large part on the reservoir's size and permeability—the characteristic measured by the rate of fluid flow through the reservoir's pore spaces.

Drive Mechanisms

There are two means by which gas and oil are driven from the reservoir; natural and man-made. The natural drive mechanisms (fig. 8–3) are as follows:

- *Dissolved gas.* As production progresses and reservoir pressure declines, gas evolves from within the oil phase and expands, forcing oil toward the lowest pressure point in the reservoir—the wellbore and the production tubing.

- *Gas cap expansion.* If the reservoir pressure is below the bubble point (its oil is undersaturated) when discovered and first produced, there will be a gas cap. With declining pressure caused by oil production, the gas cap will expand, driving the gas-oil contact in the reservoir down toward the level of the lowest pressure point—the wellbore.

- *Water drive.* If the formation in which the oil is trapped is fairly large, the water in the aquifer below the oil-water contact will expand as oil is produced, driving the contact up and forcing oil toward the wellbore.

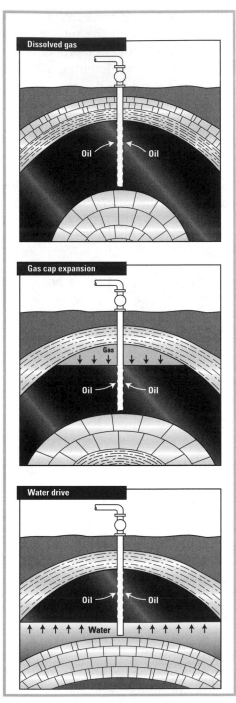

Fig. 8–3. Forms of natural drive mechanisms

Eventually, either the gas-oil (the gas cap) or the oil-water contact will reach the producing interval, requiring the operator to take remedial steps, as described in chapter 10.

Man-made drive mechanisms include water, chemical, and gas injection, as described later in this chapter, under "Secondary Recovery" and "Tertiary Recovery."

Producing Phases

The producing life of an oil well can be divided into phases, each defined by the amount of energy or pressure in the reservoir. The life of a gas well is similarly affected by pressure decline, but its phases are less complex because gas itself is a less complex medium to produce.

Figure 8–4, a plot of reservoir pressure against time, illustrates the profound importance of pressure. In the initial years of a well's producing life, reservoir pressure is high, and the fluids in the reservoir flow into the wellbore and to the surface. But inevitably, reservoir pressure declines and may drop below the pressure at the bottom of the production tubing, halting the flow. That calls for some kind of assistance.

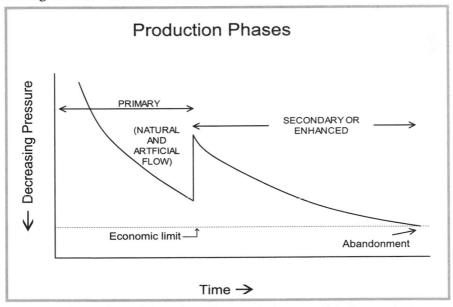

Fig. 8–4. Production phases of an oil well. As pressure in the reservoir declines, production declines. The well first flows on its own; then production must be pumped; and finally secondary recovery operations supply the energy to drive oil from the reservoir.

The first producing phase is known as the *primary phase* and includes two stages:

- *Natural flow.*

- *Artificial lift.* When the reservoir pressure has been depleted to the point that fluid entering the wellbore does not flow to the surface, the fluid must be pumped from the well or otherwise assisted in its path to the surface.

Primary Production

Natural flow

Natural flow is the simplest, most straightforward, and most profitable stage of a well's producing life. A well producing from a gas reservoir may remain in this mode for most, if not all, of its life. But with oil wells—and even some gas wells—often during the natural flow phase too much water may make its way to the wellbore. As the *water cut*—the percent water in the stream at the wellhead—increases during the natural flow phase, its increased weight (pressure) could overcome the pressure in the reservoir. (Water, with 10°API gravity, is denser than any oil that flows naturally.) While natural flow production with water cuts of up to 50% is not unusual, declining reservoir pressures make continued operation during this phase less likely.

The source of the water is often either the upward migration of the oil-water interface in the reservoir as the gas and oil are produced (fig. 8–5) or water from a waterflood project. There are remedial techniques for fixing the problem of water production, known by the operating staff as *water breakthrough*, that are covered in the next chapter. But eventually, either because of pressure decline in the reservoir or because of water, artificial lift techniques must be used to produce the remaining hydrocarbon reserves.

Artificial lift

Beam pumping. The simplest way to produce a well that no longer flows is to pump the oil or oil-water mixture out of the well. The oldest system is known as *beam pumping*, placing a pump near the bottom of the well inside the production tubing (fig. 8–6). A string of *sucker rods* (small-diameter, solid metal pipe) that extends from the pump to the surface

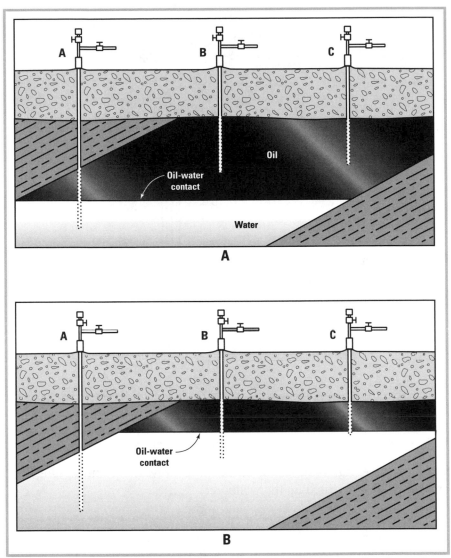

Fig. 8–5. The effect of oil production on the oil-water contact. As oil was produced, water has moved upward replacing it and watering out well A. Well B must be plugged back to reduce water production for a time, but it too will eventually water out. Well C will finally become uneconomical to produce as the water rises.

drives the pump. The sucker rod string is attached to a teeter-totter type of mechanism (a *pump jack*) at the surface (fig. 8–7), which is driven by an electric, diesel, or gas engine.

The device at the bottom of the production tubing, a *subsurface, positive displacement* pump is composed of a plunger, barrel, traveling valve, and

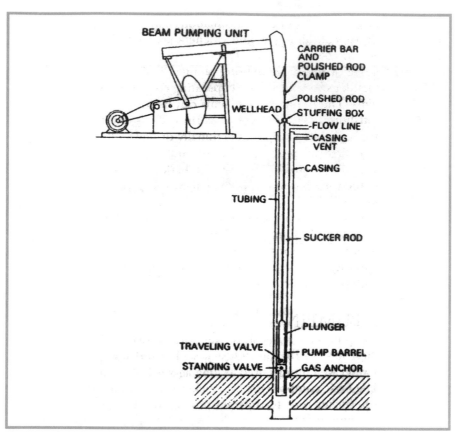

Fig. 8–6. A beam pumping unit (pump jack) operating a bottom-hole pump

standing valve (fig. 8–8). As the sucker rods draw the pump's plunger up and down, fluid is lifted up the wellbore:

- The valves in figure 8–8 open and close by fluid flow as the plunger moves up and down. At the downstroke (left) the traveling (or riding) valve opens, and the standing valve closes. The plunger moves down through the fluid, but no fluid in the wellbore moves or comes out of the well at this stage.

- At the upstroke (right), the traveling valve closes, and the standing valve opens. Fluid above the traveling valve and in the plunger is lifted up, forcing fluid out of the well. At the same time, more fluid is sucked into the well below the plunger.

Fig. 8–7. Pump jack. Near Bakersfield, California. Courtesy of Aera Energy.

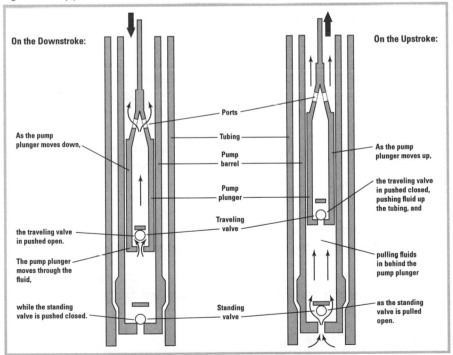

Fig. 8–8. Positive displacement pump in action. As the pump's plunger moves down, fluid in the well rises above the traveling valve. As the plunger is pulled upward, fluid above the traveling valve moves up and fluid from the reservoir is drawn into the pump's barrel awaiting the next downward stroke.

Pump jacks come in a variety of designs, but to the average citizen passing by an oil field full of pump jacks, with beams bobbing up and down, they might look like a surreal landscape of giant grasshoppers. Whatever their likeness, beam pumping units vary in size according to their service—mostly the depth and the production rate. The footprint of a pump jack unit can exceed 200 square feet.

Beam pumping units have their limitations:

- On an offshore platform, limited space prohibits even thinking about pump jacks as the driver for downhole pumps.

- In highly deviated wells, the up and down motion of the sucker rods wears holes in the production tubing.

- In deep wells, beyond about 10,000 feet, lifting the weight of the sucker rods alone requires more energy than can normally be justified economically.

- Wells pumping more than 500 barrels per day overwhelm the mechanics of most sucker rod pumps.

All these situations call for alternate means of artificial lift.

Submersible pumping. In the 1920s, the Reda Pump Company developed the *electric submersible pump* (ESP), a stationary device that sits at the bottom of a production string or is hung on its own electric line and set in fixed receptacles at the bottom of the tubing. Reda's design set the standard ever since. These pumps are most effective when large amounts of water are produced along with hydrocarbons, especially where the total fluid produced exceeds 1,000 barrels per day, as in a waterflood.

The usual configuration has the pump assembly and electric motor run on the bottom of the production tubing (fig. 8–9). An electric cable attached to the side of the tubing supplies power to the electric motor.

The electric motor develops heat as it runs and can easily run too hot and burn out. The pump is designed to pass well fluid by the motor in order to dissipate this heat.

Surprisingly, the cable supplying power to the electric motor is the weak link in the electrical system and is therefore more susceptible to failure than any other part. Frequent starting and stopping of the motor (e.g.,

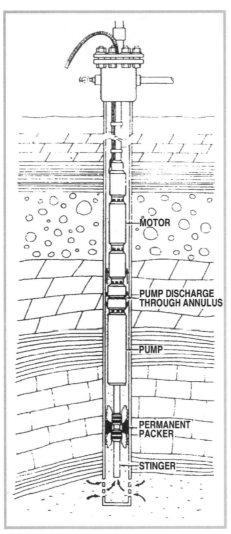

Fig. 8–9. An electric submersible pump

due to external power interruptions) drastically shortens cable life. Under such interruptible service, cable life may be only a few years.

Cable failure is due to the increase in heat developed as the electric current fluctuates. Theoretically, cable life may be as much as 10 years when the temperature reaches no higher than 167°F. Increase the temperature by 15°F, and the theoretical life declines by about half. Since temperature increases with well depth, ESPs have an effective depth limit of around 10,000 feet. That is not the measured length of the production tubing, but the true vertical depth below the surface. However, for a deeper well, say 15,000 feet, an ESP can be placed at 10,000 feet and do an effective job. Nevertheless, total recovery will be limited because the fluid column below the pump applies pressure on the formation, equivalent to the remaining 5,000-foot hydrostatic head.

ESPs are not cheap to acquire, install, or remove for maintenance or replacement, but production people are pleased to have another alternative to beam pumping.

Gas lift. Both beam pumping and ESPs push fluids up the production tubing. An alternate method, gas lift, is to inject gas into the annulus between the production tubing and the casing and then into the production tubing by a series of valves (fig. 8–10). The gas makes the column of fluid less dense—lighter—and reduces the pressure at the perforations. That lets the reservoir fluids flow in a *simulated natural flow mode*, an oxymoron but still a good description.

The only surface equipment required in order to operate a gas lift system is a compressor to raise the pressure of the injected gas to the level necessary to pass through the gas lift valves into the production tubing. The simplicity makes gas lift suitable for offshore operations, where platform space is short. Gas lift also works well in deep wells and high-angle deviated wells.

Gas lift mechanics. Shown in figure 8–10a is a well no longer flowing because the pressure in the wellbore equals the pressure in the reservoir. To remedy this with gas lift, a special production tubing string equipped with a series of gas lift valves spaced along its length is set in the wellbore. The valves are set to open and then close at successively lower pressures the deeper they are placed on the tubing. The first valve opens at the highest pressure (fig. 8–10b) and then closes after the fluid in the tubing at that point starts moving from the well.

This sequence of opening and closing and gas injection is repeated at successively lower depths until all the fluid in the tubing is moving from the well and the pressure in the wellbore is low enough for fluid in the

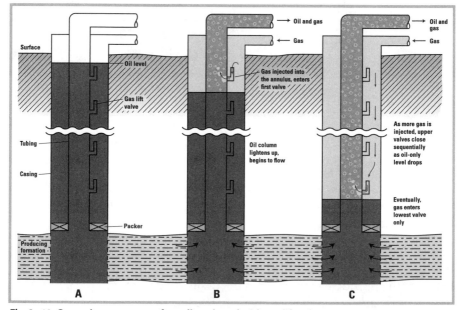

Fig. 8–10. Operating sequence of a well equipped with gas lift valves

reservoir to enter the production tubing at the perforations. Gas injection at the lowest level continues in order to keep the density of the column of fluid low enough to allow continuous flow from the reservoir (fig. 8–10c).

Gas injection can be continuous or intermittent. Continuous injection is limited to situations in which the reservoir pressure exceeds the injection pressure significantly. In the intermittent mode, gas injection is suspended for specific periods controlled by preset timing devices to allow fluid to enter the wellbore for a specified period. In this scheme, a valve at the bottom of the production string opens, gas enters the tubing to help the fluids rise, and then the valve closes, allowing more fluid to enter the tubing.

There are several other more specialized and less frequently used artificial lift systems:

Hydraulic lift. The idea of eliminating the sucker rod string and powering a bottom-hole pump by another means led to the development of hydraulic pumping. This system has four components:

- a power-fluid conditioning and supply located at the surface

- a surface power unit and hydraulic pump

- piping to transfer the hydraulic fluid from the surface

- a subsurface fluid-drive (hydraulic) pump

The fluid pump at the surface moves the power fluid down the separate piping to the subsurface hydraulic pump. The fluid drives the subsurface pump, which boosts the reservoir fluids up the production tubing. The power fluid exits the subsurface pump and commingles with the reservoir fluids.

The power fluid comes from the produced crude oil itself. To protect the pumps and motors in the hydraulic system, a portion of the produced crude is diverted and specially processed at the surface to free it from sand and other abrasive material before it reenters the system.

The large surface footprint of these facilities makes it an unlikely choice for offshore operations.

Plunger lift. This special form of gas lift is used when regular gas lift does not work because the reservoir pressure has declined to the point that very little liquid enters the tubing while gas is being injected. In this situation, a plunger or chamber is run in the production tubing with its bypass valve open. It falls to the bottom of the tubing string through any accumulated fluid. Gas is injected into the annulus and into the tubing string below the plunger/chamber, thus closing its bypass valve and simultaneously forcing the plunger/chamber to rise in the column of fluid (fig. 8–11).

The plunger/chamber and its captured fluid are forced to the top of the tubing string, where a device captures the plunger/chamber while the fluid is produced into surface processing facilities. After a designated period controlled by preset timing devices, the catcher releases the plunger; it falls to the bottom of the hole, and the cycle is repeated.

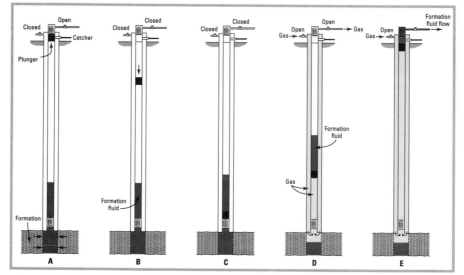

Fig. 8–11. Operating sequence of a well equipped with a plunger lift

Secondary Recovery

Reservoir pressure declines in proportion to the net volume of fluid that leaves the formation. As the hydrocarbons are produced, water fills the void left by the oil and gas, but seldom in total or even in large part. (As described in the geology discussion in chapter 3, water is found in the pore space almost everywhere in the subsurface.) While replacement is rarely complete, it does serve to slow the decline in reservoir pressure. Unfortunately, when water finally reaches the wellbore, it will begin to load the fluid column with water, which usually results in *killing* the well (causing the natural flow to cease).

Nevertheless, two of the most economical ways to deal with declining pressure and oil production involve injecting water into the oil-producing reservoir.

Pressure Maintenance and Waterflooding

Water injection into the producing reservoir is known by many names—*pressure maintenance, secondary recovery, waterflooding, supplemental recovery,* and *enhanced recovery*. In the first case, water is injected below the oil-water contact usually at the edges of the field to supplement the natural reservoir energy (pressure), to slow production decline, and to increase the ultimate amount of hydrocarbons produced. This method is also often used to dispose of produced water, especially in the offshore. In fact, produced-water disposal is by far the most important aspect of pressure-maintenance projects.

The theoretical results of the displacement of oil by water can be demonstrated in the laboratory, but the predictability of the amount by which oil recovery would be enhanced in the field usually has a wide margin of error. Two factors defy prediction—the total amount of additional oil recovery and the rate at which the recovery would occur. Both of these factors, of course, are critical to the profitability of waterflooding a field and to the future careers of the engineers recommending it.

Huge and continuing efforts to waterflood fields in the Permian Basin of West Texas, stretching from Abilene southwest toward El Paso, were initiated in the 1950s. At the time, engineers accepted the conventional wisdom that secondary recovery would equal the recovery expected from primary recovery—the ultimate recovery would double as a consequence of a full-scale waterflood. In the majority of cases it did not turn out that way, but fortunately for the participating oil companies and their engineers, oil and gas prices rose significantly in the 1970s, making the economics of waterflooding retroactively attractive.

The benefit of hindsight shows that the designers failed to understand the complexity of the reservoirs through which they wanted the water to move. The economic justification of the early floods was often based on the assumption that water would displace oil from the injection interval through the reservoir in a uniform manner both vertically and horizontally.

Such a pistonlike displacement was rarely achieved because reservoirs are seldom geologically uniform. They have stringers of fine sand lying in lenses within coarser sand bodies, fractures crisscrossing randomly, spots of relatively impermeable rock, and so on. All of this *reservoir heterogeneity* adversely affects the uniform movement of both the displacing agent (water) and the displaced fluid (oil) from the injection well toward the producing well.

Not only does the complexity of the reservoir on a micro scale undermine predictions, but water and the oil to be displaced differ drastically in characteristics between each other and from one flood to another. For instance, since water is heavier than oil, it tends to migrate to the bottom of the reservoir, perhaps missing the oil at the top as it moves forward.

The waterfloods of the 1950s and 1960s were, for the most part, modifications of existing producing oil fields. The capital required in order to convert to a waterflood was considerable, so usually existing producing wells were converted to injection, rather than drilling new wells. In large oil fields, the producing wells are drilled on a regular pattern, often one on every 40-acre plot, roughly 625 feet apart. When a waterflood is initiated, every second well in alternating rows of wells may be converted from production (fig. 8–12). The ratio of injectors to producers in this case becomes 1:1, and the pattern is called a *five spot*. Converting more wells to injection but still working on a symmetrical pattern may result in another popular configuration, the *nine spot* pattern, with a ratio of 3:1, injectors to producers.

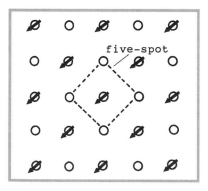

Fig. 8–12. Map view of a five-spot waterflood pattern. Every second well is an injector and every other well is a producer: one producer for every injector; but more importantly, every producer is surrounded by injection wells.

Nowadays, since most operators recognize the value of waterflooding, they space their wells with water injection patterns, either immediate or future, in mind. The critical design elements of a waterflood include

- reservoir geometry

- lithology

- reservoir depth

- porosity

- permeability

- continuity of rock properties

- fluid saturations

- fluid properties and relative permeabilities

- water source and its chemistry

Indicators of these factors come from logs, laboratory and field tests, production history, cores taken during drilling, and seismic profiling.

Since many oil reservoirs lack symmetry owing to the randomness of the formations, the ratio of injectors to producers is not always the most important design criterion. The crucial consideration is whether there is some barrier, another injection well or a geologic boundary of the reservoir on the other side of the producer, that will provide a pressure backup. Without that, the wave of injected water will simply flow around the producing well.

In more recent decades, the study of the performance of existing waterfloods has yielded much insight and potential profit enhancement. As oil prices have risen and technologies have advanced (largely seismic), it has become popular to drill between producers to tap pockets of *bypassed* oil. In some cases, the density of producing and injection patterns has been doubled through *infill drilling* an entire oil field.

Offshore, where so much of the original investment is made based on the ultimate production design of the field, waterflooding usually begins as soon as is practical. The initial producing well spacing and position are located with secondary recovery in mind.

Injectivity

The rate that water can be injected into the producing horizon at a given pressure, or the *injectivity*, is always an important but uncertain parameter. The rate of injection is directly proportional to the pressure applied to the injection stream. However, it must never exceed the overburden pressure or the *fracture gradient*. If it does, the injected water will split the rock apart, and the water will bypass oil and thus limit the success of the flood.

Laboratory tests on cores taken in the reservoir during drilling operations and examination of the injection history in waterflood projects in geologically similar reservoirs often give a clue to injectivity. Before investments are made to convert to a full-scale waterflood, pilot injection tests are often undertaken. A producing well surrounded by other producers is converted to injection. The hope is that there will be an oil response at the nearby producers, even though it is unlikely to happen on a very satisfying scale, owing to the lack of backup. However, if the injection rate remains acceptable for an adequate amount of time, perhaps several months, then investments in additional well conversions, treating facilities, and compressors may be made with confidence.

Makeup water

An essential ingredient for a waterflood is makeup water. Offshore, the water source surrounds the platform and becomes a nonissue. Onshore, produced water as a source is important from the outset and becomes even more dominant as the project progresses and the producing water cut increases. However, water production is seldom enough to replace the oil produced as well as the formation waters recovered. Before a flood is abandoned, an amount of water equal to 150–170% of the pore space in the reservoir will have been injected. This volume depends on the economics of the project, which are always very sensitive to the rising and falling price of crude.

Freshwater aquifers are often tapped. The water may be piped as much as several hundred miles to a field under waterflood. Sometimes saltwater reservoirs lying either shallower (preferably) or deeper than the producing horizon are tapped.

Ironically, produced waters require treatment before reinjection. Dissolved gasses must be removed, along with minerals and microbes. Care is taken to prevent the precipitation of minerals and the development of microbes, which tend to corrode the equipment and to plug the pore spaces in the formation. Chemicals are added continuously to the injection stream to slow the formation of precipitates and corrosion. Recovered water is kept from contact with the atmosphere to maintain an oxygen-free environment, to avoid creating a fertile incubator for the growth of bacteria.

Sweep efficiency

The efficiency of the waterflood may be rather good to astonishing on a relative scale, but there is always some residual oil left in the reservoir after flooding. *Sweep efficiency* is the term for the lateral (horizontal) and vertical degree to which floodwaters have moved through the reservoir before reaching the producing wells (fig. 8–13). Recovery of this residual oil becomes first the target of infill drilling and later the justification for tertiary or enhanced recovery.

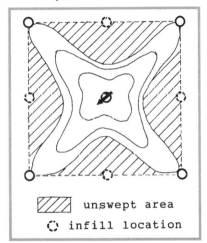

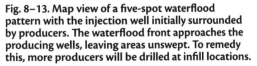

Fig. 8–13. Map view of a five-spot waterflood pattern with the injection well initially surrounded by producers. The waterflood front approaches the producing wells, leaving areas unswept. To remedy this, more producers will be drilled at infill locations.

Tertiary Recovery

Injecting water into an oil reservoir with the intent of moving oil away from the injection well, toward the producing well, is effective only when the gravity of the oil is moderate, in the range of about 17–38°API. Lower gravities (heavier or more viscous oils) do not move readily, and the injected water tends to move through and around the oil. The water arrives at the producing well having bypassed too much oil. This predictable and logical event forced production engineers to forgo waterfloods in reservoirs containing low-gravity crude oil. But, they pondered, what could make low-gravity oil behave like higher-gravity oil?

The obvious solution was to somehow reduce the viscosity of the oil, thereby increasing its mobility. Heat was the most likely viscosity-reducing agent. But how could they introduce heat into a reservoir in large enough quantities to have any worthwhile effect?

Fire floods

Producers first tried *fire floods*, pumping oxygen down wells and igniting the oil. They hoped the heat of combustion would lower the viscosity of the oil, pushing it away from the injection well, toward the production well (fig. 8–14). The pilot projects in California in the 1960s had little commercial success. The progress of the heat wave through the reservoir could hardly be controlled, and the tubulars in the producing wells failed as the heat front approached—all in all, an interesting failure.

Fig. 8–14. Cross section of a fire flood

Steam floods

In the 1960s, Shell Oil experimented with the injection of hot water and steam into heavy-oil reservoirs in Venezuela. The crews witnessed a spectacular event when the injection pressures of one well exceeded the overburden pressure. Steam, hot water, and oil erupted from a fissure created in the ground. In an anonymous stroke of insight, the crew aborted the injection operations and placed the injection well on production. Defying logic, oil flowed from the reservoir, rather than the hot fluids that had been injected.

Shell experimented with the injection of steam into shallow heavy oil (8–13°API) in California. They injected steam into the well for about a week and then produced the well for about a month before beginning the steam-injection phase again. This enhanced recovery technique was known at Shell as the "steam soak" process, but industry soon gave it the more descriptive label of the "huff and puff" technique. Steam soak eventually evolved into the continuous injection of steam in what became known as the steam drive process (fig. 8–15). Large steam generators in the field provided the steam (fig. 8–16). The two high-temperature injection processes (steam drive and steam soak) became the preferred

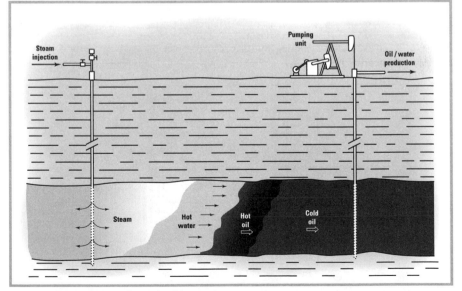

Fig. 8–15. Cross section of a steam drive project

enhanced recovery techniques for shallow (less than 3,000 foot deep) reservoirs containing heavy oil. Hundreds of millions of barrels of oil have since been and continue to be recovered from the heavy-oil fields in the southern San Joaquin Valley, near Bakersfield, California (fig. 8–17). This oil would otherwise have been left in the ground.

Fig. 8–16. Steam generators operating in a steam drive project. The Midway-Sunset Field, California. Courtesy of Aera Energy.

Fig. 8–17. Pump jacks in a steam drive project. The Bellridge Field, California. Courtesy of Aera Energy.

Enhanced versus tertiary

About the time that steam injection became a common method of improving recovery, the difference between the terms *tertiary* and *enhanced* oil recovery became blurred. Methods to add energy to the reservoir or to change the character of the contained oil (including injection of steam) came more generally known as *enhanced oil recovery* (EOR) techniques. Added to the list of tertiary or enhanced recovery techniques was the injection of solvents to clean the rock of residual oil, inert gas to provide a substitute for water injection, and carbon dioxide (CO_2). Only CO_2 injection has had any material success.

CO_2 injection

Injection of CO_2 has had wide application in the Permian Basin of West Texas and in New Mexico primarily for two reasons. First, large, naturally occurring accumulations of CO_2 lie nearby, in New Mexico and southern Colorado. Second, there are many large, mature waterfloods underway not too far from there, in the Permian basin.

Even though it is a gas, CO_2 in contact with oil at high pressures (1,500 to 2,500 psi) forms a *miscible* mixture. Like gin and vermouth, any proportion is a stable mixture, but some are better than others. As the CO_2 sweeps across the reservoir (fig. 8–18), it mixes with the oil and causes it to swell. That allows more oil to detach itself from the walls of the pores and flow to the producing well.

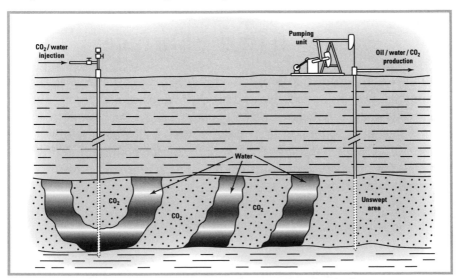

Fig. 8–18. Cross section of a WAG (water alternated with gas) project. In this case, the gas is CO_2.

Unfortunately, since CO_2 is less dense than oil or water, it tends to rise toward the top of the reservoir and bypass large quantities of oil in its path to the producing well. To combat this tendency, the CO_2 is often injected in slugs alternated with water. The process, known as *water alternated with gas* (WAG), tends to spread out the flood front, thereby increasing the sweep efficiency.

The prerequisites for a successful CO_2 project are reservoirs at depths of at least 2,500 feet (to contain the high injection pressures required in order to obtain the miscible mixture) and oil of at least 22°API. Also, most CO_2 EOR projects are conducted in a reservoir that is or has been waterflooded.

A well-designed CO_2 project can recover an additional 20% of the oil originally in place in the reservoir. Primary production plus waterflooding followed by CO_2 flooding may recover as much as 60% of the total oil in the reservoir. Sadly, the remainder remains behind, given the available technologies and today's economics.

The source for CO_2 has traditionally been natural deposits occurring in the western United States. Recently, concern for the greenhouse effects of CO_2 produced from the burning of hydrocarbons has led to the desire to sequester CO_2 generated at industrial sites and power plants into underground storage reservoirs. From there, it can be piped to appropriate oil-recovery projects.

And on to the Surface

In the early, primitive days of the oil industry, wells sometimes gushed uncontrollably to the surface. Nowadays, the hydrocarbons are too valuable and environmental concerns are too acute to allow such sloppiness. So even at start-up, production from today's wells is flowed carefully to the surface through the production tubing, into surface piping through valves and regulators meticulously adjusted to deal with rates and pressures, and on to the surface facilities.

9

Making It Possible:
Field Processing

If you can't describe what you are doing as a process,
you don't know what you're doing.

—W. Edwards Deming (1900–1993)

Hydrocarbon streams issue from the well as mixtures—some simple, some complex. Operators have to provide the surface equipment to match the complexity. These facilities separate the component parts: gas, liquids, and contaminants (mostly water, but also solids like sand and maybe some chemicals injected into the well to improve flow). Then, they have to be stored, transported, sold, reinjected, or otherwise handled. In addition, of course, the name of the game is oil and gas, and the ultimate objective of field processing is to deliver those two separately—oil to temporary storage facilities and gas to a pipeline. Along the way, the two commodities have to be made suitably clean for sale.

Some of the Parts

In the reservoir, crude oil almost always contains some dissolved methane. This natural gas usually has some liquid (or at least liquefiable) hydrocarbon components, the natural gas liquids. As production depletes the reservoir and pressure drops and as the well stream moves through various surface processing vessels, dissolved gas emanates from the oil. In a gaseous stream, liquids may condense. These phase changes are predictable once the reservoir tests covered in chapters 3 and 7 are analyzed. Equipment has to be in place to accommodate the phase changes with minimum alterations over the life cycle of the well.

Figure 9–1 shows a generalized schematic of surface operations. After the well stream exits the Christmas tree, separating it into gross components (mostly oil, mostly gas, and mostly water) is the first order of business. This is normally performed as early as is practical because it is usually easier to treat each phase separately. Besides, moving single phases—liquids and vapor—from one operation to the next requires less pressure, and that reduces backpressure on the well, which helps maintain well production.

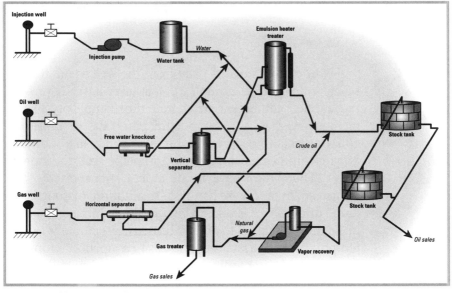

Fig. 9–1. Field processing flow diagram

Vessels designed to separate phases only—vapor from liquid—are known as two-phase separators. Three-phase separators are used when free water in the production stream also requires removal. Separators come in a variety of designs, sizes, and shapes and are most often known by their shape: vertical (fig. 9–2), horizontal (fig. 9–3), or spherical. Spherical separators are the least popular of the three because, although compact in form, they have limited separation space.

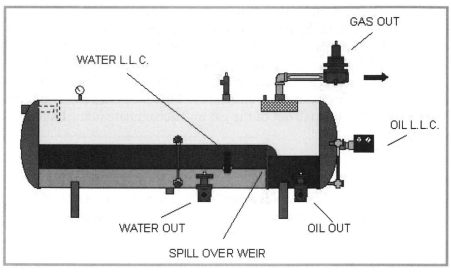

GAS OUT

WATER L.L.C.

OIL L.L.C.

WATER OUT

OIL OUT

SPILL OVER WEIR

Fig. 9–2. Vertical separator. L.L.C. is liquid level control. Courtesy of United Process Systems.

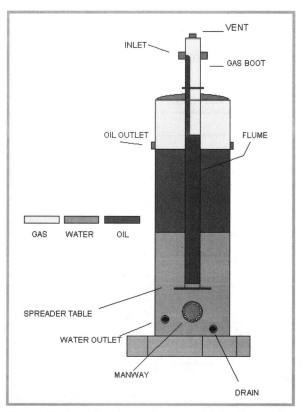

VENT

INLET

GAS BOOT

OIL OUTLET

FLUME

GAS   WATER   OIL

SPREADER TABLE

WATER OUTLET

MANWAY

DRAIN

Fig. 9–3. Horizontal separator. Courtesy of United Process Systems.

In any of them, the bulk of the liquid is separated from the gas by baffling at the inlet of the separator. The baffles create centrifugal forces in the vertical separators or an abrupt change of direction in the horizontal separators. Gravity carries the gas up and the liquids down. The stream is further separated in the next section of the vessel, where additional baffles reduce the gas velocity and the turbulence. That facilitates the formation of droplets of oil that fall out of the gas and accumulate in the bottom of the vessel.

A mist-extraction section near the top of the vertical separator or near the far end of the horizontal separator collects additional droplets of liquid. The accumulated liquids at the bottom of either unit exit the vessel, controlled by a level-control valve. Separator pressure is usually controlled by a backpressure-regulating valve on the gas outlet line.

Horizontal and vertical separators each have their advantages and disadvantages, as table 9–1 shows. Vertical separators are typically used when the gas-oil ratio is very low. Horizontal separators are preferred when the volumes of gas or liquids are large.

Downstream of the separator, the production streams go through their own processing schemes.

Table 9–1. The pros and cons of separators

Vertical	Horizontal
Pro	**Pro**
More versatile—handles a wide range of gas/oil/water compositions	Smaller diameter needed, smaller size overall
Small foot print	Less expensive
Liquid level control is less critical	Easy to skid mount and transport
Handles sand, scale, and paraffin better	The dispersion mechanism reduces foam and turbulence, increasing gas-oil separation
Con	**Con**
Larger diameter needed	Less versatile
More expensive	Larger footprint
Difficult to skid mount	Liquid level control is more critical Difficult to clean out sand, scale and paraffin

Gas Treating

Gas conditioning or *processing* takes gas separated from the well stream and creates a commodity safe for storage, transportation, and sale. Contracts for sale and transfer from the lease into a pipeline specify (among other things) the delivery pressure. In the early life of a gas field, the pressure is normally higher than the minimum pipeline requirement, which is often around 1,100 psi in the United States. As the field pressure declines, compressors must be put into service, first to enable delivery and second to assure that the field wells are not restricted in their ability to produce because of pressure held against them. In fact, in some of the oldest fields, the producing pressure is so low that compressors are put in place to draw a vacuum on the wells, to coax the gas out of the reservoir. The compressors have to do extra duty to get from a negative pressure in the wellbore to pipeline requirements.

Acid gas

Other pipeline specifications include maximum amounts of water, inert gases (nitrogen being the most common), and acid gases (hydrogen sulfide [H_2S] and carbon dioxide [CO_2]). Inert and acid gases both lower the heating value, and the acid gases cause corrosion in pipelines and other equipment. Hydrogen sulfide is also a deadly, toxic pollutant.

Natural gas containing no H_2S is called *sweet gas.* A gas with any appreciable amount of it is known as *sour gas.* Pipelines generally call for a maximum of four parts per million or 0.25 grams per standard cubic foot. There are dozens of processes for recovering sulfur compounds from natural gas. The most prevalent for onshore use are the *batch* and *amine* processes.

In the batch process, the sour gas passes across some reactant, like ferric oxide, in the form of iron filings. The hydrogen sulfide reacts with the iron, forming iron sulfide and a small amount of water. Eventually, the iron sulfide builds up, and the "spent" iron oxide has to be changed out. Because of this, the batch process is used where the gas flow rate and the hydrogen sulfide amounts are low, and the initial low equipment cost justifies the expense of discarding the reactant.

Liquid amines can be used in high-flow-rate cases for which the high capital cost of the equipment can be justified by the lower operating costs. In the amine process, the sour gas goes into the bottom of a treating vessel full of baffles. *Liquid diethanolamine* (DEA) or a similar amine compound is fed to the top of the treater. As the so-called *lean* DEA sloshes past the gas, it absorbs the H_2S, allowing the gas to exit from the top of the treater as sweet gas. The so-called *fat* DEA comes out the bottom of the vessel and goes to a heater where the H_2S flashes out easily. The cleaned-up amine is recycled to the treater. The H_2S goes to a converter where it is turned into a disposable waste product.

When the sulfur content is below 20 pounds per day in the gas stream, the batch process is usually the most economical. When the sulfur content exceeds 100 pounds per day, the amine process is the choice. Between the two values, other considerations such as dehydration needs may tip the scale.

The other so-called acid gas, CO_2, is found in varying quantities in produced gas streams. CO_2 has zero heating value, so excess amounts, more than a couple percent, reduce the sales value of the gas. Amine treaters remove CO_2, as well as H_2S, and can be put into operation where CO_2 is a problem.

Water

Free water in the form of salt brine is sometimes produced along with gas in the well stream. Low temperatures and high pressures are ripe conditions for free water and natural gas to form *hydrates*, a slushy mass that can plug flow lines. Their control and abatement is a principle concern in production of natural gas onshore and especially offshore, where the gas may travel a long distance in a pipe in cold conditions along the seabed before it is brought to the top of a platform. Remedies include keeping the gas warm, above the hydrate dew point, by any of the following methods:

- Heating the line with steam or electricity

- Insulating the line, including:

  - using double-walled pipe with a vacuum between the walls, as is done in some deepwater operations

  - wrapping the pipe with insulating material

- injecting additives like methanol into the gas to lower the freezing point

Free water aside, all natural gas is saturated with water vapor at the temperature and pressure conditions of the separator. (Water that condenses downstream of the wellhead or in the gas-oil separator condenses as essentially fresh, rather than salt, brine.) Pipelines generally require a water content of a gas stream to be no more than 7 lb/MMcf (pounds per million cubic feet). To obtain this degree of "dryness," the gas stream can be treated in a glycol treater or in a desiccant or an adsorption system.

A glycol treater (fig. 9–4) operates much like an amine treater. Gas enters the bottom of a vessel with baffles or *packing* (loosely packed filler).

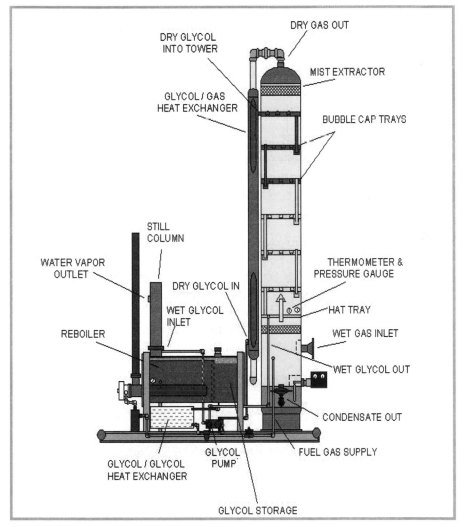

Fig. 9–4. Glycol treater

Liquid *triethlylene glycol* (TEG), or a similar compound from the family of chemicals used to make automotive radiator coolant, enters the top of the vessel. As the TEG sloshes past the gas, it absorbs the water, allowing the water-free gas to exit through the top of the treater. The TEG is moved to a heater, where the water evaporates and is released. The TEG, now rejuvenated, is recycled to the treater.

Another alternative, *adsorption systems,* is simpler and less expensive to buy but more costly to operate. These treaters have a compound, an *adsorbent,* in the main vessel that attracts water but not hydrocarbons. The water molecules stick to the adsorbent's surface by molecular attraction, which is why they are called adsorbents and not absorbents, which would hold the selected substance *internally*.

Desiccants like activated charcoal are a cheap form of adsorbent. Molecular sieves are a more sophisticated, more effective, and higher-cost version. Molecular sieves are specially manufactured, tiny metal particles with holes, measured in molecular dimensions, in their surface. The holes are constructed to allow only certain-sized molecules, in this case, water, to enter and be retained. As with the desiccants, the water has to be driven off after all the voids are about full. For this reason, adsorption systems generally have parallel systems so that one system can be adsorbing while the other is drying out.

A dry desiccant system is useful for gas flows up to 200 MMcf/d (million cubic feet per day). For higher volumes and in places where space is an issue, the glycol and molecular-sieve systems are more common. Back in figure 9–1, a schematic flow diagram showed the production streams processed for the removal of oil, water, condensate, and contaminates in order that they could be sold from the field and transported by pipeline.

Oil Treating

Oil must be "clean" to be saleable. It must meet certain buyer specifications for sediment and water content. The oil patch terms these contaminants *BS&W* (basic sediment and water) and sometimes, more simply, *S&W* (sediment and water). Standard limits are 0.1–3.0% by weight.

Methods to remove sand, salt, water, sediments, and other contaminants come under the name *oil treating*. Fundamental to oil treating is using gravity to separate contaminants from the oil.

Water is the largest-volume contaminant in either oil or gas production. In some cases, it exceeds 10 barrels of water per barrel of oil (a water-oil ratio [WOR] of 10), especially in older fields. In many situations, much of this water comes to the surface as an emulsion, which annoyingly will not separate in standard separators (fig. 9–5).

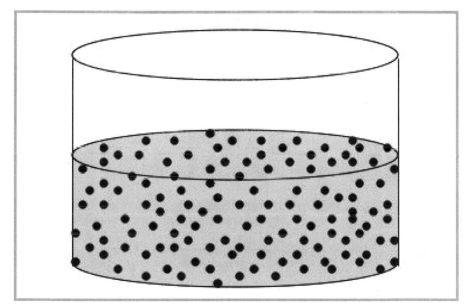

Fig. 9–5. Emulsion

Oil-water emulsions can be broken at the lease by any of four methods:

- increasing the settling time to allow the droplets to fall out of their own accord

- heating the emulsion

- applying electricity

- adding demulsifying chemicals

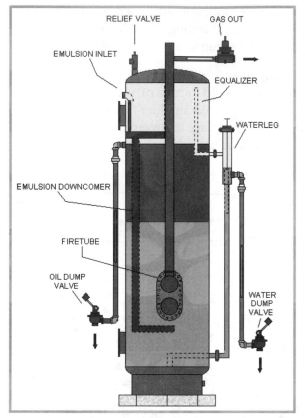

Fig. 9–6. Heater treater. Courtesy of United Process Systems.

Heater treater

The devices used to add heat to the well stream to break oil-water emulsions come in two shapes, vertical and horizontal, with a firebox at the bottom (fig. 9–6). The oil-water emulsion contacts the firebox directly to pick up the heat. At higher temperatures, the attraction of oil to water declines, and the water droplets settle out more rapidly. The fuel for the firebox is either demulsified crude or natural gas. In the case of crude, it is fed by a small pipe from the body of the treater to the burner tips. Natural gas can be taken from any place in the gas system after the gas has been separated from any liquids and is clean enough for sale.

Free-water knockout vessel

Heater treaters cannot handle much free water—vertical heater treaters are more limited than horizontal ones. Putting a *free-water knockout vessel* (FWKO) upstream of the heater treater does the job (fig. 9–7).

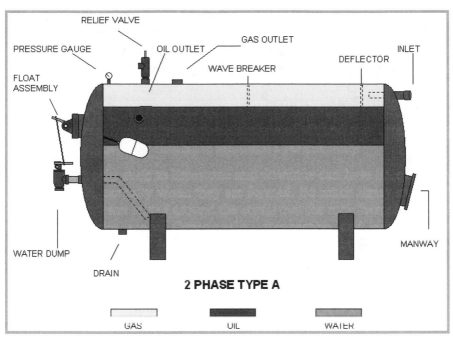

Fig. 9–7. Horizontal free-water knockout (FWKO). Courtesy of United Process Systems.

The size of the FWKO depends largely on the volume of fluids passing through it and the residence time to separate them. Free water separates from the well stream in 10–20 minutes. A well stream of 1,500 barrels per day calls for an FWKO with a volume of about 50 barrels. To provide an extra margin of residence time, most FWKOs have a volume of 400 barrels or more.

Another popular form of FWKO is a *gun barrel* (fig. 9–8), so called because of the vertical tube, the barrel, located in the middle of the tank. The tank is kept about one-third full of water. The production stream enters the top of the tube and flows to nearly the bottom of the tank, well below

Fig. 9–8. Typical oil field processing facilities. From the left, a heater treater, a gun barrel FWKO, three oil storage tanks, and a large water storage tank with an injection pump and injection well in the right foreground.

the oil-water interface, where it exits the tube, rises across a spreader plate, and separates. The oil rises to the top of the tank, and the water remains behind. Float valves maintain constant water and oil volumes in the tank, with the excess amounts going to their respective storage tanks.

Electrostatic heater treater

When water droplets in an emulsion are very small, their weight allows them to settle only very slowly. However, water molecules are polar—they carry an electric charge. In an electric field, the small droplets are attracted to each other. They coalesce into larger droplets and consequently separate more rapidly from the oil.

Electrostatic heater treaters are similar to standard heaters except that they contain electrodes in the coalescing (quiet) section. Electrostatic treaters are more compact than heater treaters, making them the preferred device where space is at a minimum and a large volume of crude must be treated—as on offshore platforms, urban areas, and arctic environments, where equipment sits on platforms over the tundra. In addition, electric dehydration breaks down emulsions at temperatures that are only 20–40°F. That saves fuel costs and reduces crude shrinkage and gravity loss. (At lower processing temperatures, less gas liquids boil out of the crude, leaving more, lighter crude.)

Choosing among combinations of FWKOs, heater treaters, and electrostatic dehydrators is a trade-off between capital and operating costs. The electric units are more expensive at the outset, but operating cost and size restrictions sometimes make them the treater of choice.

Demulsifying agents

Vendors sell a number of proprietary, complex chemicals to break emulsions. The tiny size of the oil droplets in an emulsion makes oil's normal electric repulsion of the water ineffective. Emulsifiers generally weaken or destroy the stabilizing film surrounding the dispersed oil droplets. Adding a small amount of heat along with the demulsifying chemicals to the mixture permits the dispersed water droplets to join other droplets and fall out of the oil by gravity.

Normally, demulsifying agents are added continuously to the oil stream by a small pump. Like so many other pieces of auxiliary equipment, the pumps, the injection rate of the demulsifier, and the quality of the treated oil stream have to be checked at least daily.

Water Disposal

Water is separated from the production stream at the separator, at the FWKO, at the heater treater or electrostatic treater, and at every possible opportunity during processing. If the water is to be injected into a subsurface formation either for disposal or as part of an enhanced oil recovery (EOR) project, it may have to be treated before injection. If not, the formation—and consequently, the disposal or EOR project—can be seriously impaired.

After separation from the oil and gas streams, the produced water stream may still contain harmful quantities of

- suspended solids—removed by settling in tanks, by filtration, and with hydrocyclone tanks where gravity settling is accelerated by spinning the fluid

- suspended oil—removed by allowing the oil to float in skimmer tanks and by hydrocyclone tanks

- scale particles—controlled and eliminated by continuous injection of scale inhibitors

- bacteriological matter—treated with either inorganic oxidizing agents (chlorine or sodium hypochlorite) or organic biocides (oxygen or nitrogen compounds)

- corrosive acid gases (H_2S, CO_2)—controlled by using sacrificial anodes, using corrosion-resistant materials in piping and tanks, and adding corrosion-inhibiting chemicals to the water stream

Produced water is often pumped into a well specially drilled for reinjection into the reservoir (fig. 9–8). Besides disposing of the sometimes large volumes, with no other cheap outlet, the water helps maintain the reservoir pressure or complements a waterflood underway.

A reinjection well can be too expensive to drill because of the depth of the reservoir or the geology of the formation. If that is the case offshore, the water may have to be treated to remove contaminants in order to qualify for a permit to discharge it into the sea. Onshore, water may have to be trucked or pipelined to a certified water-disposal facility, always an expensive proposition. The cost of water disposal varies from negligible to oppressive, depending on the site and the method—from a per-barrel cost from five cents to several dollars.

Testing

When a well is ready to produce for the first time, the production staff has an estimate of what will come out—oil, gas, or a combination. Still, they want to know how much and under what pressure, to better understand the reservoir's mechanics and to design the facilities needed to process the production.

An initial well test, the *potential test*, provides a best estimate of the maximum daily producing capability of the well under a fixed set of circumstances. Under most circumstances onshore, a gas well may be flowed with no restrictions and with the production being flared. The time period of the test depends on the circumstances, but concern for wasting a valuable resource and abiding by emission permits generally restricts a flare test to a few hours. The flow rate and pressure performance are carefully measured, and the results are converted mathematically into what is called an *open-flow potential test*. This gives the rate, expressed in MMcf/d, at which the well will produce without backpressure from the surface equipment.

For an oil well, the initial test has more issues. Flaring the entire stream is usually not a feasible or even a legal option. Instead, a portable test unit consisting of a separator, meters, and storage tanks are brought to the well site for the test. The gas is flared; the water and oil are separated and stored for later removal. The length of the test, usually numbered in hours, is limited by available storage.

Once permanent processing and storage facilities have been built, wells are individually tested through the facilities on a daily or weekly basis. Normally, at least one well at a field production unit is being tested all the time.

Offshore, pollution concerns and physical facilities limit the scope of initial well tests, a source of much frustration for explorers who want to test a discovery well. Instead, limited tests are made with wireline tools that capture appallingly small quantities of fluid and measure pressure performance over extremely short periods. Production staff extrapolate this scanty information to potential daily rates under given sets of pressures. Companies invest huge sums of money on the basis of this limited information, so they rely on their experienced staff to interpret this meager data very accurately.

Measurement and Metering

Hydrocarbons are valued commodities. At the point when custody or ownership of the hydrocarbon is transferred, the volume being turned over to a second party has to be known precisely, to determine payment. In addition, it is important to know about volumes at other points along the production stream, to accommodate engineering and operating needs. Production engineers need to know the daily flow rates and the volume of water produced, to determine reservoir performance and operating efficiency. Accordingly, production is measured at a minimum of two locations—the wellhead and the point of custody transfer. Since the flow at the wellhead contains a combination of oil, gas, water, and maybe sand and other trash, the content has to be periodically sampled and analyzed for composition. Measurement at the point of ownership change is to assure proper payment.

Gas metering

Gas flow can be measured using either an *orifice* or a *turbine* meter. Both are in widespread use.

Orifice meters. The centerpiece of an orifice meter is a constriction in the gas flow line. Using a theory developed by Venturi and Bernoulli several centuries ago, plus the properties and pressure readings from upstream and downstream of the constriction, the flow rate is calculated—or more accurately, *inferred*, since the volume is not measured directly. The equations for converting the raw data to flow rates have been developed over the years by American Petroleum Institute committees and provide standards acceptable to buyers and sellers.

An orifice meter, known as the *primary element*, consists of an orifice plate, housing, two pressure sensors (one upstream of the orifice plate, one downstream), and a temperature sensor. The readings from the pressure and the temperature sensors go to a *secondary element* to be recorded. At older, smaller leases, the secondary element might have recordings on paper charts (fig. 9–9). In newer, larger operations, the secondary element is a computer that can manipulate the data into flow readings.

Since gas is easily compressible, the mass passing by the meter is a function of pressure, temperature, and composition. All three factors go into the calculation of the flow. In many cases, the composition of the gas

Fig. 9–9. Gas well meter. The volume of gas sold is calculated from pressure recorded on paper charts like that shown here. The charts are collected regularly by the lease operator and read by technicians at the company office.

varies enough that the two parties agree to have a *densitometer* (a device for measuring density) track the gas quality and be included in the flow calculation as a variable, instead of a constant.

Turbine meters. Like orifice meters, these devices do not measure volumes directly. They measure the velocity of the fluid passing through them. A turbine meter has a windmill-like device in the flow line. The faster the flow rate is, the faster the blades rotate. The speed of the flow is directly proportional to volume, but not mass. Like an orifice meter, a turbine meter used for gas flow has to have pressure and temperature sensors and perhaps a densitometer to make appropriate adjustments for the compressibility and composition of the gas. Turbine meters can be used to measure flow rates from 250 Mcf/d to 3.5 MMcf/d.

Oil metering

Oil has such different properties from gas that the measurement techniques differ substantially. The facilities at a lease for the transfer of oil depend in part on the vintage of the facility and the volume to be measured. At old, small production leases where oil is hauled away in tank trucks, volumes can be measured by meter at the loading rack or by measuring the change in the volume in the stock tank. The meter of choice is generally a *positive displacement* (PD) meter.

Positive displacement meter. These devices measure the actual volume of fluid passing through the line by isolating an exact amount with a piston, the revolution of a gear, a sliding vane, or some other mechanical engineer's clever invention. The meter, again the primary element, transmits a count to a mechanical or electronic counter.

Alternatively, the lease operator, acting as a gauger, can climb to the top of the stock tank to be unloaded and drop a measuring tape into it. After the truck has received its load, the gauger repeats the measuring-tape drop. By pencil or by portable computer and using volumetric tables for that particular stock tank, the loader can get the total volume loaded and transferred to the second party.

In either case, the gauger takes samples of the oil to determine the S&W, either at the site or later.

Turbine Meter. Turbine meters are used for oil measurement, as well as gas. In oil service, they tend to be more finicky than PD meters at lease locations. They work best with lighter fluids, such as condensates. Heavier crudes are tough on the turbines that drive the system. In between the very light and the heavy crudes, the choice is mixed. In any event, the turbine meters have to be calibrated periodically with an independent meter prover to assure their integrity.

Lease automatic custody transfer (LACT). At more progressive leases, where oil continuously transfers into a pipeline, LACT units perform the measurements (fig. 9–10). The central focus of the LACT unit is still the PD meter, but a handful of events take place continually and automatically before the oil reaches it:

- A strainer removes solids.

- A pump may increase the pressure to pipeline requirements.

- An S&W probe checks the oil, particularly for water content. Typically, it uses electric current to measure the dielectric constant (the capacitance) of the oil. Higher water content has a lower dielectric constant. Exceeding a preset value causes the oil flow to be diverted to lease treating facilities or to just shut down.

- A de-aerator removes excess, non–revenue-generating air and protects the meter from slugs of air.

- A diverter valve is actuated by the S&W probe.

- A thermometer records the temperature.

- An automatic sampler with a storage container retains part of the sales stream for later analysis.

- A PD meter measures the fluid volume.

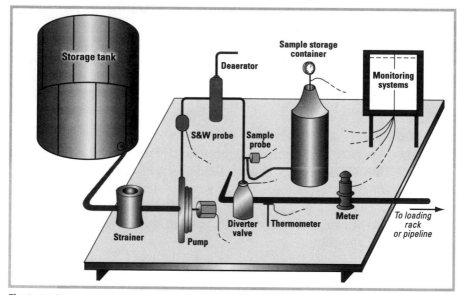

Fig. 9–10. Lease automatic custody transfer (LACT) unit

Storage

In the early days of the oil patch, storage tanks were not much more than giant barrels made of wooden staves, held together by steel straps. The wood absorbed the fluid in the tank, causing it to swell and seal, reducing—but often not eliminating—leaks. By the 20th century, riveted and then welded steel tanks had proliferated.

High schoolers learn that the volume of a cylinder (a tank) is found by multiplying its height times its horizontal area. They also (should) know before they graduate that the circumference of a circle is found by multiplying the radius of the circle by two times *pi*. Thus, with a little algebra, they should be able to figure out the same thing that *strapping crews* do when the calibrate field tanks.

Despite their faultless appearance, tanks are never perfect cylinders, regardless of the effort of the construction company to make them look that way. To calibrate a stock tank, the strapping crew measures the circumference of the tank at multiple levels, normally every five feet. From the circumference at each level, they can back-calculate the tank radius and then the area at that level. That gives enough information to extrapolate the tank's volume every quarter inch and to create a tank table unique for the particular stock tank, which shows the volume increments from any one liquid level to any other liquid level.

Onshore, when a stock tank has been sufficiently filled to warrant shipping and the method is to transport the crude oil by truck rather than pipeline it to a refinery, it is *sealed* and *gauged* to determine its volume. The lease operator (or pumper) will divert all additional production to another stock tank and then drop a tape (a measuring device) into the full tank and determine how full it is. How many barrels it contains is determined with the tank tables. The lease operator will arrange for a tank truck to pick up the crude. The truck driver will make his own measurement of the tank and then hook up to the tank's offloading line and open the valve letting the oil empty into his truck. The driver will then gauge the tank again to determine the volume that he has collected. Later, the lease operator will gauge the tank to verify the volume shipped. In this fashion, both the seller (employer of the lease operator) and the buyer (employer of the tank truck) have independent measures of the volume transferred from seller to buyer.

If delivery of the crude is into a pipeline, then a LACT unit and its meters is the method of measuring the shipment. LACT units are independently certified periodically with meter provers to satisfy both seller and buyer of accurate measurements.

Offshore, production is metered directly into a pipeline or in some instances, the oil is moved temporarily into the tanks of an *FPSO* (a floating production, storage, and off-loading vessel; fig. 9–11) or an adjacent *FSO* (floating storage and off-loading vessel). In either case, the oil awaits a shuttle tanker to offload the oil and take it to market. Determining the precise size of the compartments on any of these vessels has to be done by certified meters. The oil is measured, usually by turbine meters, as it is transferred to the shuttle tanker.

Fig. 9–11. An FPSO (floating production, storage, and off-loading vessel). Courtesy of Veritas DGC.

And On . . .

In the next chapter, discover what happens when things go wrong in the producing well—and going wrong is an almost assured occurrence.

10

Fixing Things: Remedial Operations and Workovers

Nobody realizes that some people expend a lot of energy merely to be normal.

—Albert Camus (1913–1960)

Murphy must have been thinking about oil and gas wells when he coined his third law, "Anything that can go wrong will go wrong." There are few wells, facilities, platforms, or oil field projects that do not have their share of problems and failures. Seldom can these events go unattended, for they deal with recovery of valuable assets in the ground and the return on the large sums of money invested to find and produce them. This chapter deals just with producing wells. They are the means by which the hydrocarbon assets are accessed, and the remedies to their problems are complex enough to fill books. (This book only covers the most common ones.)

The chamber of producing-well horrors has in it

- Mechanical failures: a sucker rod string may break, valves in the sucker rod pump may become worn, a submersible pump may wear out, corrosion may cause a hole in the tubing or casing, gas lift valves may need replacement, and packers may need repair or replacement.

- Producing characteristics may change: an increase in water production is common, and often production rates decrease drastically.

- Plugging may occur in the production tubing or across the producing interval either within the well or in the reservoir.

- Injection pressure may increase dramatically because of plugging of the formation due to scale formation, causing the water injection rate to fall off.

- A general disappointment may result because any of Murphy's rules and the local situation can cause the well to perform below its potential.

Decision Making

In many instances, the operating staff feel compelled to move forward with remedial operations at the slightest indication of a problem. After all, their mission is to maintain production and injection targets. But the decision must always be based on economic principles. Often, it is intuitively obvious that the cost of remedial operations (often called *well intervention*) is far below the potential gain from getting the well back on production, and repairs can proceed right away. But too often the job is so obviously complex, burdened by risk, and potentially so costly that an analysis of the economics is required before proceeding.

After determining what needs to be done to solve the problem at hand and the manner in which the work is to be performed, the staff estimate the cost to remedy the situation. On the other side of the ledger is the incremental revenue resulting from the production rate increase above what it would have been without intervention. That amount is multiplied by the expected future price of oil (or gas or both); then, the workover cost and future operating costs are subtracted to determine the economic attractiveness of proceeding.

Without remedial work, oil/gas recovery from this point forward can range from zero if the well is *dead* (not producing) to a significant amount, but always less than what it might be with a proper workover. In many instances, future production can be calculated from graphical procedures (fig. 10–1).

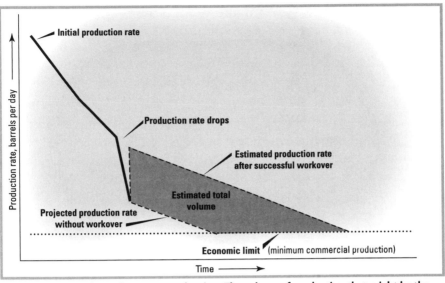

Fig. 10–1. Impact of a workover on production. The volume of production that might be the result of a successful workover can be calculated from the difference between the two curves.

Workover Rigs

Normally, moving in a well-servicing (workover) rig is the first step in repairing a well problem. Such a rig is much like a drilling rig, yet smaller and far more portable (fig. 10–2). Generally, there is no need for the drilling rig's mud system of pumps and tanks, and often there is no need for elaborate pressure-control equipment. The workover rig's primary function is to remove the sucker rod string and pump or the production tubing from the well. The workover rig also can provide a mast for running wireline clean-out and repair devices into either the tubing or casing.

Before pulling the tubing or doing most other operations on the well, it must be *killed*—that is, production must be stopped, and the pressure in the reservoir must be contained. This is accomplished by unseating any tubing-casing packers and then circulating fluids down the tubing-casing annulus and up the tubing string. The weight of these fluids balances the pressure in the reservoir and prevents the influx of any reservoir fluids. Where the reservoir pressure is relatively high, weight materials (often ground-up calcium chloride [$CaCl_2$] or limestone, i.e., calcium carbonate [$CaCO_3$]) can be added to the kill fluid. Unfortunately, this material has a tendency to be forced back into the reservoir, reducing its permeability and requiring an additional step and perhaps the additional cost of stimulation (covered below) in the workover operation to remove the offending material.

Fig. 10–2. Workover rig. All land based workover rigs are mounted on a motorized trailer. When ready to be moved to another job, the top half of the mast (derrick) is telescoped into the larger, lower half and the entire assembly is tilted forward to horizontal.

Once the well's pressure is under control, the tubing can be removed or devices run on wireline can be used to fix the problem. Wireline tools include bailers to remove sand or to place small amounts of cement in plug-back operations, packers, perforating guns, flow-rate sensors, and of course electric logging sondes.

Coiled Tubing

An alternative solution to pulling the tubing is to employ *coiled tubing* (CT). This system allows the operator to enter the well against pressure and to perform remedial activities inside the tubing under a variety of circumstances, both onshore and offshore. One benefit of using CT is the elimination of the time (cost) of pulling the production tubing. The tubing itself is a continuous length of steel or composite tubing, flexible enough to be coiled on a large reel for transportation. The diameter depends on the intended application and usually ranges from $\frac{3}{4}$ inch to $4\frac{1}{2}$ inches. The reel can range from 13 to 20 feet in diameter.

A CT unit (fig. 10–3) consists of the reel of coiled tubing, an injector, a control console, a power supply, and a well-control stack through which the uncoiled tubing is injected. Ancillary equipment usually needed includes storage tanks for liquids and pumps to circulate fluids into the

Fig. 10–3. CT operation. Flexible tubing is taken from the reel and "injected" into the well through its existing production string. The power for the injection process is supplied by the hydraulic lines that extend from the trailer to the platform on the right of the photo. Note the series of blowout preventers stacked on top of the christmas tree (the assembly with the wheel controlled valves). Courtesy of BJ Coiltech

well. If foam is injected as the circulating medium instead of liquid, the equipment to generate and inject it is needed.

CT can be run into the well much faster and more cheaply than can be conventional, jointed tubing. CT has no joints to screw together and needs no well-servicing rig to deal with tubular goods. In addition, fluids or foams can be circulated continuously through the tubing while it is being run into or out of the well. These two features—running time and the ability to operate without killing the well and perhaps damaging the formation—can translate into huge cost savings. These are the drivers, of course, behind the popularity of using CT in workover operations.

Subsea Completions

Wells that have been completed on the ocean floor present perhaps the most expensive and complex well-intervention situations. These *subsea completions* include a submarine wellhead and a subsea tree. Traditionally, a mobile offshore drilling unit, perhaps a small semisubmersible drilling vessel, is required in order to put in place and support a workover riser extending from the surface of the vessel to the subsea tree. The riser acts as an extension of the casing. Included in the riser package are blowout preventers, emergency disconnect devices, circulating hoses, and control modules. Wireline or CT can then be run into and down the riser and into the well. From that point forward, the workover operation assumes a more conventional mode.

Well Problems

Mechanical failures
All of the moving parts in the well—such as pumps, sucker rod strings, and gas lift valves—are exposed to continuous wear and tear. Even tubing is worn by sand abrasion and by the reciprocating sucker rod string. Tubing and casing can both corrode to the point of failure. Tubing can be replaced; casing patches can be conveyed into the well by tubing, CT, or wireline and can be set across the failed interval of the casing. Casing failures can also be repaired by squeezing cement into the problem interval.

Water production

Initially, the well may produce only oil or gas or a mixture of the two. More often than not, there is a water-bearing interval either below the perforations or in close lateral proximity. When water production becomes a problem, the standard solution is to *plug back* with cement—that is, place cement across the lower portion of the perforations, hoping to exclude the interval contributing a majority, if not all, of the water (fig. 10–4). This is a temporary solution and may have to be repeated a number of times, each in a higher interval, until the well *waters out*—that is, until the water-oil ratio becomes so high that it is uneconomic to produce from this interval.

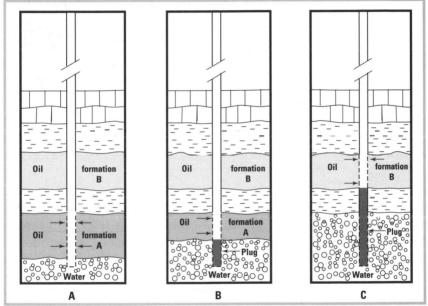

Fig. 10–4. Remedial operations to reduce water production. In A, the casing has been perforated above the oil-water contact and only oil is produced. In B, the interface has risen as oil is produced and the water bearing part of the reservoir has been plugged off with cement. In C, the water has risen to the top of the formation requiring that the well be plugged. Perforations have been shot opposite the oil bearing portion of the next higher formation and oil production resumed.

Coning

Another instance of unexpected water production can occur through a phenomenon known as *coning*. Sometimes an oil or gas zone lies directly on water—there is an oil-water or gas-water interface within the productive interval penetrated by the well. Usually, every attempt is made to perforate as far from the water zone as possible. Occasionally, a combination of high production rate and high permeability causes enough of a pressure drop around the wellbore for water to be drawn upward to the perforations (fig. 10–5). The only solution is to reduce the production rate in the hope

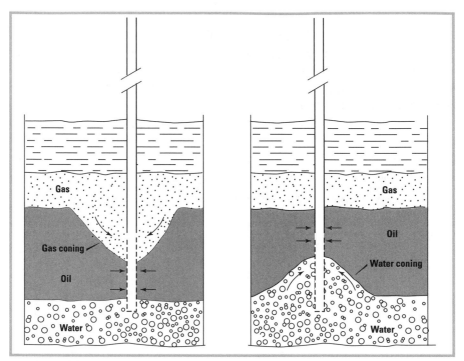

Fig. 10–5. Gas and water coning

that the cone will collapse and the amount of water produced will decrease. This is a very discouraging situation for the operator who has based the economic justification for drilling the well on a much higher production rate than can be sustained without coning.

Coning of *gas* downward into an oil-producing interval can also occur, but the results are less disastrous because all of the production can be sold. However, coning gas into the underlying oil zone accelerates depletion of the gas and reduces the benefits gained from gas cap expansion drive, thereby reducing overall oil recovery. Here again, not much can be done other than to reduce the production rate.

Plugging

Many reservoirs are composed of loosely consolidated sand. This sand has a tendency to move from the formation, through the perforations in the casing, and into the wellbore. When this is expected, the initial completion of the well can include a gravel pack or a screen intended to prevent sand influx.

These solutions are not foolproof, and the sand may invade the wellbore anyway. The sand can cover the producing interval and can be carried up into the tubing, plugging it and slowing or even stopping production. Removal of sand plugging in low-pressure wells is generally accomplished by removing the tubing and bailing the sand from the producing interval. In higher-pressure wells and almost always offshore, the sand is circulated from the well by using CT. Circulating fluid is injected down the tubing and up the tubing-casing annulus with enough velocity to carry the sand out of the well. It can also be "reverse circulated"—down the tubing-casing annulus and up the CT. This technique has the advantage of an increased fluid velocity (upward in the CT) and is a more effective method of removing sand.

Plugging of the tubing can also be caused by the precipitation of *scale* (mineral deposits—most often some form of carbonate or sulfate). Scale can plug tubulars, gravel packs, perforations, and even the formation itself. It occurs in both production and injection wells and is the result of changes of temperature and pressure in the produced water or the mixing of incompatible waters in injection wells. Scale problems are one of the most serious problems associated with producing operations.

One solution is placing acid across the area where scale has formed. This works to a degree with carbonate scales, but sulfate scales are less soluble in acid and are difficult to remove chemically. As an alternative, CT with mills and carbide brushes on the bottom of the string can be run inside the tubing to clear it of scale. Often, scaled-up perforations must be reperforated.

Precipitation of paraffin in the production tubing and surface piping can also cause plugging. Many crude oils contain dissolved paraffin in the reservoir. As the oil moves up the tubing and cools below the melting temperature of the dissolved paraffin (about 124°F), it deposits as a solid wax on the interior walls of the tubing. This is a common occurrence offshore, where the tubing is cooled by cold seawater. Onshore, the usual solution is to dissolve the paraffin by circulating heated crude oil down the casing and up the tubing. This may be required periodically during the life of a well. In wells equipped with sucker rod pumps, the sucker rod string can be equipped with scrappers attached to the sucker rods, which mechanically keep the tubing walls clean as the rod string is reciprocated up and down.

Offshore, insulated tubing strings may prevent deposition below the surface. Also, heated water can be circulated within a special tubing string run alongside the production tubing, warming the paraffin like a heat exchanger.

Well Stimulation

It is well established that production rates in all wells decline as the energy (pressure) in the reservoir is depleted by production. The falloff of this rate is predictable, and each well establishes a decline character that it follows throughout its life unless it is affected by any number of extraneous events not associated with pressure decline. Besides the problems already covered—mechanical failures, water intrusion, perforation, or tubing or interval plugging—the formation itself may lose its original permeability. Fine material may move into the vicinity of the wellbore, or scale may precipitate near the face of the producing formation. A solution can be to *stimulate* the well with either of two basic methods, *acidizing* and *fracturing*.

Acidizing

The most benevolent form of stimulation is the acid job. Any one of numerous acids is placed across the producing interval and allowed to dissolve the offending material from the producing formation. Care must be taken that the dissolved material is not displaced (pushed) back into the formation a few feet, re-creating a plugging situation. Keeping the dissolved material in the wellbore can be accomplished by maintaining the hydrostatic pressure during the acid job at less than the reservoir pressure and by circulating the spent acid and the dissolved minerals from the well without delay.

Fracturing

Fracturing the formation is in reality splitting the rock by using fluids and high pressure to create a near-vertical fracture that may extend several hundred feet in two directions, at 180° from each other, away from the well. This phenomenon is successful because fluid flows from the boundaries of the reservoir toward the wellbore. The nearer each molecule gets to the well, the more crowded the flow channels become. High-permeability reservoirs may handle this easily. In tight reservoirs where the permeability is low, fluid flow may be constrained initially to such an extent that the production rate is not commercial without fracturing.

Fracture stimulation is most effective when applied to hard rock as opposed to soft sand. The producing formation must be solid enough to split, rather than to be squeezed like a sponge. For example, production in the Gulf of Mexico and onshore near the coastline is mainly from unconsolidated sands where fracturing finds limited application. In contrast, in the interior of the United States, the producing formations are hard rock—limestones and sandstones that lend themselves to fracture stimulation.

Often the *frac treatment* is performed as part of the initial completion procedure. This occurs when it is considered likely that the well may not

Time, Risk, and Rates of Return

The value of future production has to be reduced by the consideration of the time value of money. One dollar received some time in the future is worth less than one dollar received today. Most of the value of the production stream will be received in the future.

Calculating this present value uses a *discount rate* that is related to the cost to a company of the money it invests. Some companies use their borrowing rate. Others use their *cost of capital*, a blend of their borrowing cost and the cost of their equity (stock).

If either is, for example, 8%, the value of $100 received in five years is only $68 today. The value of a stream produced over five years that delivers $100,000 profit at the time it is produced may be worth only $82,000 today, depending on the production rate decline and the discount rate.

One way companies assess the attractiveness of any investment outlay is to compare it to the present value of the future return. They divide the discounted present value of the profit by the investment dollars required in order to achieve it. If the resulting ratio (the reward ratio) exceeds a company's threshold—many use 15%—the investment in remedial work gets approved.

The internal rate of return (IRR) is another popular evaluation method. It is the discount rate that makes the future net cash-in equal to the present cash outlay.

The risk that the remedial job will fail can also be considered. The team of engineers and operations people might mutually agree that the chance of success of the operation may only be three out of four, or 75%. To account for this risk, they can reduce the expected return by 0.75 before they calculate the profit-to-investment ratio or IRR.

produce at an economical rate unless stimulated at the outset. That is the case with coal-bed methane producing intervals and other *tight* formations that have little natural permeability but contain economic reserves of gas or oil.

Rock splits if pressure is applied to a localized area, such as at a perforation or series of perforations in the casing opposite the producing interval. Although the fractures move laterally, normally the split is vertical. It is easier for the rock to be pushed apart sideways than it is to lift the entire weight of the overlying rock strata.

Normally, but not always, shale formations lying above and below a producing interval confine the vertical fracture within the intended formation. But unless there is movable water either above or below the producing zone, containing the fracture within the formation only affects the efficiency of the treatment. (Creating a longer rather than a higher fracture likely contacts more of the producing interval.)

In theory, fracturing allows the wellbore to be affected by a greater amount of the reservoir. A fracture changes the well's open access from less than one foot to tens and perhaps hundreds of feet. The effect on the production rate may be an increase of 5–50 times.

Fracing (pronounced "fracking") operations. Fracturing the formation in the lingo of the oil patch is to perform a *frac job*. Typically, as many as two dozen service company pump trucks are driven to the well site and hooked up to the wellhead by a series of pipes and manifolds. Portable tanks are also trucked in to hold the liquid, which is pumped against the formation to create the fracture. Along with the pump trucks and tanks are trucks carrying *propants* (fig. 10–6).

The frac fluids normally used include low-gravity oil, water mixed with a variety of chemicals to thicken it, or water carrying dissolved carbon dioxide or nitrogen. Propants consist of small, uniform sand grains or equally sized plastic pellets that are mixed with the frac fluid and carried by it into the fracture.

The fracture is created by exerting all the pumping pressure of the pump trucks through the surface piping and manifold, down the tubing and against the formation. On charts, at the surface is shown the gradual rise in pump pressure to the point at which the fracture gradient is exceeded.

Fig. 10–6. Well site with truck-mounted pumps tied together. In this way, high pressure can be obtained to create a large fracture in target formation.

Then there is an abrupt drop in pressure, indicating that the formation has been fractured.

Once the fracture is created and begins to propagate away from the wellbore, it is held open by the pressure created by the pump trucks' pumping fluid into the well. But as soon as the pumps are shut down, the frac fluid tends to bleed off into the formation, allowing the fracture to heal completely. To hold it open, propants or propping agents are injected along with the fracturing fluid. When the pressure is released, the propants remain in place, holding the sides of the fracture apart.

Acid is used instead of a propant when the producing formation to be fractured is a limestone or dolomite, formations easily dissolved by acid. The acid is injected at the front of the frac fluid. It partially dissolves the faces of the fracture, creating irregular, etched intervals through which fluid or gas can flow after the pressure is released and the fracture closes.

A well may be fracture stimulated several times during its producing life. This is often because the propping agents have ultimately been crushed by rock pressure or have migrated back to the wellbore, allowing the fracture to close.

Changing Production Intervals

Many wells penetrate several *pay* intervals representing potential, multiple producing horizons. Each has its own production character controlled by formation characteristics such as porosity, permeability, water saturation, and often pressure. Normally, the productive intervals are kept separate on the initial completion by the cement placed between the casing and the borehole. Generally, the lowest interval is perforated and produced to depletion before the next-higher interval is produced.

Onshore, when the lowest interval is depleted, the tubing is pulled, cement is squeezed into the producing interval, the next-higher producing interval is perforated, the tubing is rerun, and the well is returned to production. Offshore, where all workover and recompletion operations are so much more expensive than onshore, the initial completion may provide for subsequent plug backs and completions without removal of the producing string. In such cases, the casing is perforated opposite several potentially productive intervals. They are separated by packers run on the outside of the tubing that seal the tubing-casing annulus between pay intervals. The tubing is equipped with sliding-sleeve devices placed

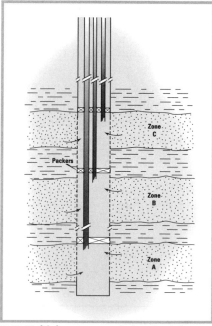

Fig. 10–7. Multiple-zone completion. A triple completion: three zones producing simultaneously but separately. Each is produced through its individual tubing sting and packer.

opposite the future intervals to be produced. On the tubing between these interval are placed landing collars. When it is apparent that the initial producing interval is to be abandoned, a plug is run inside the tubing and set in the landing collar above the depleted interval and below the sliding sleeve opposite the intended new interval. That sleeve is then opened, and production can begin. Alas, too frequently, the sleeve fails to open because of corrosion or grit or some other aspect of Murphy's law. Then, the far more expensive operation of removing the tubing and proceeding with a normal recompletion must be performed.

As another alternative, the tubing may be set above the highest interval that may be produced. The casing at the lowest pay interval is perforated and produced to depletion. Then, it is capped with cement and/or excluded below a packer, and the next-higher pay interval is perforated. All this is performed with wireline devices run through the tubing string.

A third alternative is to complete as many as three pay intervals, but separate them with packers and simultaneously produce each up its own tubing string (fig. 10 7). Sometimes one interval fails during production, requiring workover. In general, little attempt is made to restore production in that zone until the other zones experience difficulty requiring intervention. Too much valuable, ongoing production would otherwise be lost.

Abandonment

After a well has been produced to depletion or has failed beyond repair, it must be abandoned. Such a wanton name, *abandonment*, belies the strict protocol required by the government agency with authority over the area.

The normal procedure onshore is to remove all surface equipment and production tubing, place cement plugs across the producing interval(s), recover all uncemented casing, place cement plugs opposite all freshwater intervals, cut off the surface casing several feet below the ground surface, cap the well, and level and generally tidy up the location before leaving. The procedures are similar to a well intervention, with the exception of pulling of the uncemented casing strings.

Onshore, companies specializing in casing recovery often take over the abandonment operation at no cost to the operating company. They profit from the salvage value of the recovered casing. Hydraulic jacks are used to apply a strain to the particular casing string, and a *free point indicator* is run on a wireline to determine where the casing is cemented and where it is uncemented, or free (to move). A jet cutting device is run in the hole on a wireline, and the casing is cut just above where it has been shown to be free. It can then be jacked and hoisted out of the hole and hauled to the service company's storage yard for eventual sale. Usually the recovered casing is too corroded or worn to be used in another well, but it still finds its way into numerous alternate uses—surface line pipe, cattle guards, road culverts, and the like.

Occasionally, a well no longer capable of commercial production is retained for possible future use perhaps as an injector or observation well in a waterflood. Such a status is called *temporary abandonment*. In order to meet most agency requirements, pressure on the casing annulus and the production tubing must be frequently monitored and reported.

Offshore, many successful exploratory wells are left in such a mode that they can be reentered and placed on production when appropriate surface and subsea facilities have been constructed. They are, in fact, temporarily abandoned. Otherwise, the offshore exploratory well is permanently abandoned by placing cement plugs opposite potential producing intervals, placing a cement plug in the surface casing, and cutting off everything a few feet below the mud line (the seafloor).

And On . . .

The topics surrounding the production of oil and gas are done. Next come the questions of who the people and organizations are who carry out this work.

11

Who's Involved:
The Players

*In a little time I felt something alive moving on my left leg; bending
my eyes downwards as much as I could, I perceived it to be a
human creature not six inches high.*

—Jonathan Swift (1667–1745), *Gulliver's Travels*

The Companies

The U.S. oil and gas industry bears close resemblance
to Lilliput, a land of a giant (or two) and thousands of
tiny people. That is not meant to disparage one group
or another in the industry. Companies of all sizes make
many people in them wealthy; small companies produce,
as a group, more rich people than do large companies. At
the same time, all companies and their stakeholders—the
owners, the investors, and the employees—face potentially
debilitating risks. The consequences of that are covered in
the next chapter.

Size
What are the facts? The U.S. Department of Commerce
statistics tell the story. More than 6,000 enterprises produce
oil and gas in the United States. Shown in figure 11–1 are
only the largest 50, arranged along the horizontal axis in
order of decreasing size. The vertical bars measure their
individual production rates against the left scale (in barrels

of oil equivalent, or *boe*, a combination of oil and gas in one volumetric term). The curve of the cumulative production of the 50 producers, starting with the largest, is measured against the right scale.

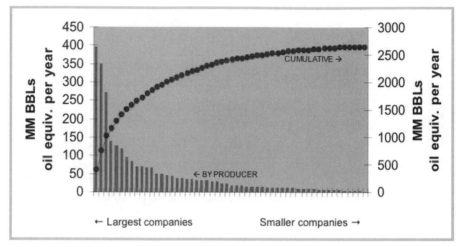

Fig. 11–1. U.S. oil and gas production by the 50 largest producers. Out of about 6000 companies, the 50 largest account for about half the U.S. production.

In decade after decade, wave upon wave of acquisitions in the industry—some would say consolidation—certainly has left behind behemoths. As figure 11–1 shows, the top 20 producers extract from the ground about 50% of the country's petroleum. The 50th-largest company produces only about 11,000 boe per day. And beyond that, the size of the company falls off, and the remaining 50% of production comes from the other 99% of the producers.

The 99% are relatively modest in size indeed. They each produce, on average, 2,500 boe per day. The U.S. Bureau of Commerce reports that 70% of all oil- and gas-producing companies have less than 10 employees. By almost any measure, the U.S. oil and gas industry has not consolidated.

Beyond the United States and a few other, mostly English-speaking countries, the structure of the oil and gas industry is quite different. Whereas the United States has a large number of small companies and relatively few large companies, most other countries have only a small number of very large companies. Their structures vary and overlap:

- a national oil company that has exclusive rights to oil and gas operations, such as Pemex in Mexico

- a national oil company that partners with international companies in many, if not all, oil and gas ventures in the country, such as Petrobras in Brazil and Nigerian National Petroleum Company (NNPC) in Nigeria

- a quasi-national oil company that competes, often in an advantaged position, with international companies for oil and gas operations, such as Statoil in Norway

Composition

Small companies do not always emulate large integrated exploration and production (E&P) companies. Specialization divides players, not surprisingly, by geographic focus but also, not so evidently, by function. In the thousands of smaller companies, various business models have long been active, including

- classic companies that explore and then produce what they find

- companies that specialize in exploration and sell what they discover at an early stage of development (often with overrides, a small share of the revenue from future production, despite not having ownership)

- companies that acquire producing properties—some quite mature, some quite immature—and operate them

- companies that own but do not operate, leaving that to other owner-operator partners or a company that provides operating services

- investors that have a financial stake but no operating role in production

The People

Toiling in the companies are people of various titles and academic training. From the outside, the designations seem bewildering. From the inside, there is often intense pride of academic disciple. Anyone who has mistakenly called a geophysicist a geologist knows that.

In a large company, a team working on a proposed drilling prospect located onshore will include a combination of individuals with technical and professional backgrounds on the one hand and operations skills on the other. Included might be a geologist, a geophysicist, a drilling engineer, a drilling foreman, a petrophysical engineer, a landman, permitting staff, a reservoir engineer, a production engineer, an operations foreman, a facilities engineer, regulatory staff, and tax staff.

Clearly, a small company does not have on board the luxury of all that talent. They have to outsource some expertise. Other inputs have to be covered by whoever is available and has the skill to contribute. But what do all these people bring to the table anyway?

Geologists

These scientists are trained in their specialty at universities. Seldom do they rise to a sufficient level of expertise by the "school of hard knocks" alone and thus avoid university training in their specific academic discipline as do many of their professional colleagues, the engineers. Geologists' interest in their academic discipline, plus a liking or disliking for details, generally leads them in a career path toward either exploring or producing projects.

Exploration seems to require a more expansive and intuitive orientation, while production projects require attention to detail and extrapolation of those details into working hypotheses. Exploration geologists bring to the party knowledge of Earth mechanics (plate tectonics), and sedimentary systems, as well as hydrocarbon migration and accumulation. Production geologists are more deeply familiar with lithology, reservoir boundaries, and locations of faults and have an understanding of the history of deposition that created the target reservoirs. They take the lead in

- distinguishing reservoir rock from nonreservoir rock

- defining the dimensional continuity and the character and the gross thickness trends of the reservoir

- providing a framework for specifying reservoir rock porosity, permeability, and capillary properties

- estimating how much pore space there is, how much hydrocarbon fills it, and how it might flow

There are specialty assignments held by geologists who are expected to contribute within a narrow (but deep—no pun intended) range of needs. Among these are

- *Stratigraphers*—scientists generally in the exploration part of a company who focus on the unique character of particular geologic horizons and correlate these from one locale to another; often providing prediction of the depth at and circumstances under which an horizon will be encountered during exploration drilling.

- *Paleontologists*—geologists who have studied the evolution of ancient organisms. In the industry, these scientists identify the foraminifera contained within sedimentary layers with the purpose of correlating one layer with another, thereby helping other geologists unravel the mysteries of what has been or may be encountered in the subsurface—the effort called making maps.

Engineers

The geologist's professional colleagues are the engineers. Most have a technical degree. They may have specialized in one of the engineering disciplines (mechanical, chemical, petroleum, electrical, etc.) or in one of the sciences, like chemistry or physics. Whatever their background, many engineers in large and small oil companies began their careers in larger companies that trained them in the discipline of petroleum engineering. That schooling often covered or complemented the courses taught in university petroleum engineering schools, often to a level equivalent to an advanced engineering degree.

The trained, so-called "petroleum" engineers lose their academic identity and assume any of numerous industry engineering roles, from operations engineer and drilling engineer to reservoir engineer or petrophysical engineer. Often their career path leads them to become an operations or drilling superintendent before achieving higher management levels.

Trained engineers or geologists are mobile and marketable, and many leave the large companies that trained them and take positions with smaller firms, where they contribute in numerous engineering and operations activities. In essence, they have to don whatever professional hat the current project requires and exploit the breadth of their past experiences and training.

Engineering titles

Regardless of their academic training and prior industry experience, engineers are often titled in accordance with their current assignment. The contribution of each has to be couched in the broadest terms and not in any particular order of importance:

- *Reservoir engineer*—deals with the theory and application of fluid flow in porous media (oil, water, and gas moving in the reservoir); generally, is responsible for determining the theoretical size of an accumulation using test data, its economic value, and (along with the geologist) the design of systems to recover the hydrocarbons.

- *Petrophysical engineer*—has an understanding of the fabric of rocks and of wireline log analysis; often a geologist exposed to and trained in logging or an engineer similarly trained; is responsible for interpreting electric logs and advising on pay intervals and the amount of hydrocarbons contained in each, and is often asked to predict anomalous pressures in a drilling project and interpret logs for indications of changes in the pressure while a well is being drilled.

- *Drilling engineer*—a mechanically oriented engineer who enjoys operations, rather than technical work; expected to play a key role in the design of a well to be drilled and to contribute expertise during drilling.

- *Production engineer*—another mechanically oriented engineer with broad exposure to well problems and producing techniques; is expected to design production systems and remedial operations to improve production; often, in the absence of a facilities engineer, designs and organizes maintenance of field processing equipment.

- *Facilities engineer*—brings an understanding of the chemistry of petroleum and its products, the manners in which it may be treated to meet sales quality; often has a background in chemical engineering or chemistry; works with vendors to design and obtain the proper vessels to treat production. This role is often outsourced to an experienced service/supply/fabrication company.

- *Structural engineers*—a specialist, generally having a degree in mechanical engineering and usually employed either by a fabrication company that designs and manufactures offshore platforms and their components or by a drilling company in the design of their rigs. Large E&P companies generally have their own structural engineers to interface with the contractors.

Geophysicist

Like the engineer, the geophysicist is often the product of training by a large company or a specialty service company. Academically, the profession is a combination of geology, physics, computer science, and electrical engineering. There are three distinct needs in industry: seismic data acquisition, processing, and interpretation. An interest in geologic concepts—either exploration or production oriented—plus an appreciation for the physics of the Earth leads many geologists to become geophysicists. Training in physics and/or electrical engineering may prejudice some toward a career in data acquisition. Degreed computer science graduates gravitate into the field of seismic data processing. Interpretation is generally the realm of the geologists and geophysicists.

Operations staff

These employees implement the designs and plans of the technical staff. Among their administrators are the *superintendents* of drilling or production. These usually have an engineering background or are employees who have shown unusual promise in the field and are trained in directing and otherwise administering field activities. The groups divide into two, the hands around while wells are being drilled and those who operate the wells when they come onstream.

Drilling. The people who provide drilling services for the most part work for drilling companies, not the oil and gas companies. The crews include

- *Rig manager*—so named for the obvious reason, but in the field called the *pusher* or, in Canada, *the push*. Onshore, a rig crew amounts to four to six hands; offshore, the number may swell to a dozen or more, depending on the complexity of the operation.

- *Driller*—in charge of the remainder of the crew and operator of the rig; needs good mechanical aptitude and an understanding of the operation of all of the components of the rig; almost always will have been promoted from the ranks of the crew.

- *Derrick man*—works in the derrick, standing drill pipe back in the derrick or pulling it out, depending on whether coming out or going in the hole (making a trip); when not so employed and while the rig is drilling, the derrick man makes sure the mud system is maintained at design specifications.

- *Roughneck*—one of several crew members working on the floor of the rig, making drill pipe connections during trips; while drilling, will use the hoisting apparatus to pick up additional joints of drill pipe (fig. 11–2).

- *Roustabout*—the lowest rung on the ladder of the driller's crew and sometimes called *lease hand*; will be assigned miscellaneous tasks such as painting, removing thread protectors from drill pipe, moving sacks of mud chemicals into and from storage, and just about anything else that needs doing that can be done by a novice.

Fig. 11–2. A roughneck takes a break

Production. The people dealing with producing operations work either for the oil and gas company or a contract production company. They include

- *Production foreman*—may be an engineer or field employee who has been promoted from the ranks; usually is assigned control over an entire field, several fields, or a large part of a single field, depending on the size and complexity of the operation. Offshore, the production foreman may be responsible for the entire production operation of one or two platforms. The production foreman's staff includes, in the simplest case, pumpers (oil well operations) or gaugers (gas well operations). These may be employees of the company. More often they are contract personnel, engaged by the operating company or outsourced to another company providing this type of service. Collectively, they are also known as lease operators.

- *Pumpers/gaugers*—responsible for keeping wells producing, advising if they stop producing, changing measurement charts, locating leaks, and so forth.

- *Roustabouts*—have a place in production as well drilling operations; supply the muscle to perform repairs under the direction of the production foreman. Often, they are supplied by a service company.

In addition to these hands, an army of craftspeople provide all sorts of construction and maintenance services—electricians, mechanics, plumbers, pipe fitters, instrument technicians, carpenters, welders, metalworkers, drivers, and more. Some work for the operators, but in most locations, the majority are employed by service companies.

The order in which the above occupations are listed is not intended to be a hierarchy of talent. Without meaning to be patronizing, people in all these categories have remarkable degrees of proficiency. Some, by circumstance or choice, start and continue their careers as craftspeople. Others rise through supervisory ranks and perform competitively next to engineering school graduates. Collar color has never been a good indicator of talent or intelligence in the oil patch.

Supporting staff. These may be either outsourced employees or members of the E&P company.

- *Landman*—usually a college graduate who has the responsibilities of first determining who has title to land the company wants to drill on and of then negotiating the lease with the landowner or mineral owner. Most companies have a landman or at least have a close association with a company doing this business. After drilling and during the producing life of the well (field), the landman will act as the interface between the landowner and/or the mineral owner and the operating company.

- *Lawyer*—only the larger companies employ a full-time lawyer. Legal matters are generally outsourced. They may range from regulatory matters to tax interpretations to workers'-compensation problems.

- *Accountants*—these degreed financial professionals have many roles in any type of company. Unique to oil and gas companies is the disbursement of income to royalty owners, partners, and investors, as well as joint interest billing (invoicing partners for their shares of the cost of operations—usually capital investment).

The Suppliers

Oil and gas production today would not be possible without the contributions of thousands of oil field service companies. The breadth of their input defies summary, except for the iceberg cartoon in figure 11–3. In addition, the center of gravity of innovation and technological development has long since shifted from the E&P companies to the service companies.

One way to measure the importance of the service industry is to use the sage advice from Watergate's Deep Throat: "Follow the money." How much spending by E&P companies goes to their own employees for wages, salaries, and benefits, and how much goes to outsourcing?

E&P companies do not report their outside expenditures directly, but combing through their filings with the Securities and Exchange Commission allows a good estimate. Since many companies in the industry have activities other than exploration and production—power

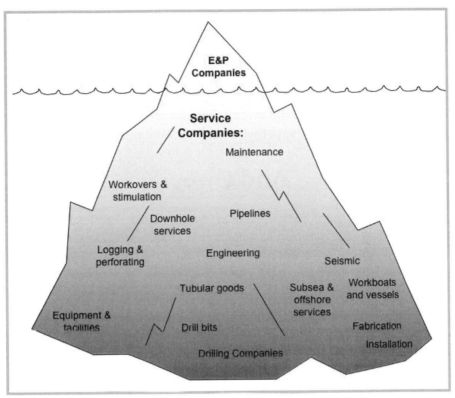

Fig. 11–3. Distribution of effort in E&P. Most of the dollars and manpower expended in E&P are in the service industry, not the oil and gas companies.

plants, regulated pipelines, trading businesses, and other distractions—it is helpful to look just at nearly pure E&P companies. Those tend to be large independents, and they form a reasonable sample.

Looking at recent performance—taking the expenses for operations, plus general and administrative expenses plus the capital expenditures for increasing the values of operating properties—gives an approximation of how much companies spend on business activity. That includes all the drilling, equipment purchases, oil field services, construction and installation, and other activities involved with exploration, development, and production.

Companies also report how many employees they have. With an educated guess at what the total salary and benefits per employee come to, the total internal spending can be deduced. Comparison of the spending on employees and total spending is startling. For the dozen or so largest independents, only 5–15% of E&P spending goes to employees and their benefits packages. The rest goes to purchase goods and services from outside companies. Some 85–95% of E&P spending is outsourced to oil field service companies.

Clearly, the thousands of small and large companies in the oil field service sector that cater to the exploration and production industry are the sine qua non—without them, there would be nothing.

And On . . .

If 85–95% of the money is spent elsewhere, what *is* an E&P company anyway? That's the question addressed in the next—and last—chapter.

12

What Should We Do?
Strategy

The lion and the calf shall lie down together
... but the calf won't get much sleep.

—Woody Allen (1935–), *Without Feathers*

Companies do not survive just by deciding they want to be oil and gas producers. The number of producers in the United States declined in the last two decades of the 20th century, in round numbers, from 8,700 to 6,200 (fig. 12–1). Staring in the face of that mortality rate, how does an oil and gas company choose a path to success?

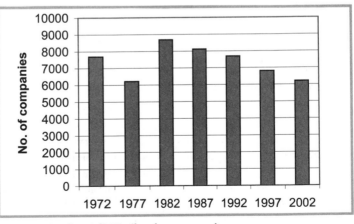

Fig. 12–1. Numbers of U.S. oil and gas companies
Source: U.S. Department of Census data

Identity

What is an oil and gas company? This book has covered hundreds of activities essential to producing oil and gas. But that is not to say that an E&P company (or department or division of an E&P company) needs the personnel and skills to do them all. To the contrary, over the history of the oil and gas industry, most activities have migrated out of the oil and gas companies to service companies.

Thirty years ago, many E&P companies, large and small, performed the processes shown on the left in figure 12–2. But even at that time, most, especially the smaller companies, outsourced specific, specialized services such as drilling, seismic surveys, workover operations, and wireline services.

By the end of the 20th century, service companies had positioned themselves to provide most of the activities needed by the upstream oil industry. Many oil and gas companies had eagerly outsourced activities traditionally considered their exclusive province—even operations,

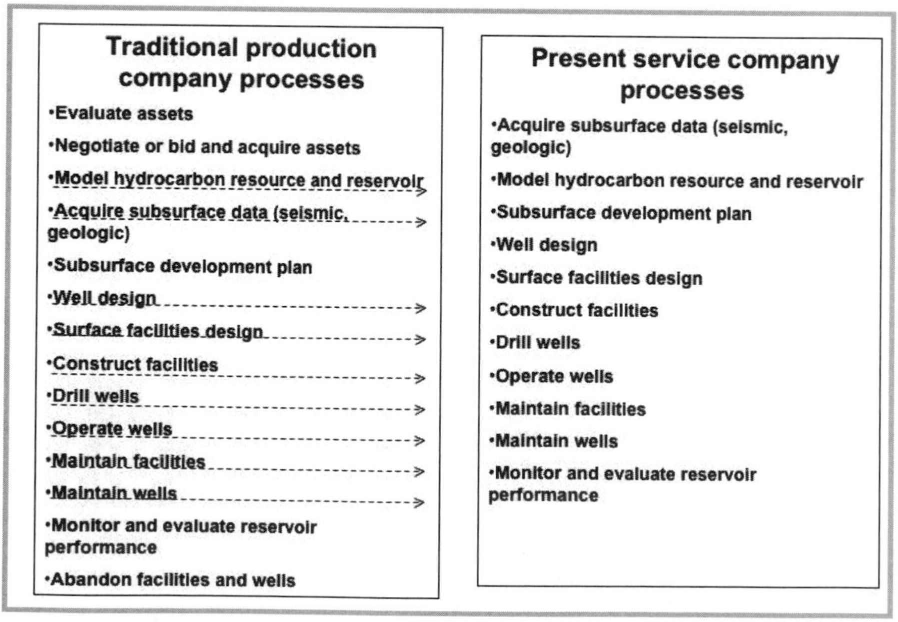

Traditional production company processes	Present service company processes
•Evaluate assets	•Acquire subsurface data (seismic, geologic)
•Negotiate or bid and acquire assets	•Model hydrocarbon resource and reservoir
•Model hydrocarbon resource and reservoir	•Subsurface development plan
•Acquire subsurface data (seismic, geologic)	•Well design
•Subsurface development plan	•Surface facilities design
•Well design	•Construct facilities
•Surface facilities design	•Drill wells
•Construct facilities	•Operate wells
•Drill wells	•Maintain facilities
•Operate wells	•Maintain wells
•Maintain facilities	•Monitor and evaluate reservoir performance
•Maintain wells	
•Monitor and evaluate reservoir performance	
•Abandon facilities and wells	

Fig. 12–2. Migration of production processes from E&P companies to service companies

maintenance, and subsurface monitoring. The company, Total N.A. reported using 42 different service companies to establish the production at their Matterhorn platform in the Gulf of Mexico.

So what *does* an E&P company do? Sit a chief executive officer and a chief financial officer in a loft somewhere with a computer, a list of e-mail addresses, and a checkbook—a virtual oil company hiring third parties to do whatever? That hardly sounds like a path to a rewardingly successful—or even acceptable—future.

A Strategy Model

One approach to a sustainable strategy comes from adapting the competency framework devised a few years ago by two specialists in strategy. Their research convinced them that successful companies had in their strategies three vital attributes (fig. 12–3):

- Technological innovation—the ability to continuously deploy cutting-edge techniques to business processes. Some companies lead in developing innovations in-house and applying them. Some work closely with innovative vendors and take up their technologies quickly. Either way, they get an advantage over competitors in numerous ways (e.g., initiating frontier production, producing hard-to-produce hydrocarbons, increasing recovery or production rates) where others cannot.

Fig. 12–3. Attributes of clear, successful strategies

- Operational effectiveness—some companies wring costs out of operations like water from a wet rag, giving them more profitability and more cash flow to reinvest than their competitors.

- Intimate relationships—some companies have honed their market acumen. They understand their business environment, the world external to them, much better than others. They get access to and shuck assets because they know

  - which competitors are weak and vulnerable to takeover

  - how various governments and their agencies and others who control access work

  - what financing entities (internal or external) require

  - which competitors will pay more for their own assets than they themselves can value them

The degree of competence in each of these categories can range from outstanding (even world-class) to competitive (as good as any other company working in the same area) to noncompetitive (unable to perform as well as the average). Based on the observations of the two strategy specialists, *sustained success and competitive advantage is more likely achieved when a company is outstanding in at least one category (seldom in all three) and at least competitive in the others.*

Some generic (but successful) E&P strategies emerge from the three-legged model (fig. 12–4):

- Strategy 1. *First Movers* rely on technological innovation to tie up assets before competitors. Examples include Shell Oil, who explored the Gulf of Mexico Outer Continental Shelf faster than anyone, then leapt into the deepwater on the basis of their understanding of the unique geology; and Petrobras' pioneering use of floating production platforms in the deepwater of offshore Brazil

- Strategy 2. *Fast Followers* take up others' technological break-throughs quickly and combine them with cost-effectiveness to establish a sustainable position. Examples are British Petroleum, who watched (and even partnered with) Shell as they progressed in the deepwater of the Gulf of Mexico. They quickly applied Shell's world-class production techniques to their own deepwater discoveries around the world.

- Strategy 3. *Scavengers* combine market acumen (understanding weak competitors) and cost-effectiveness to value properties higher than existing owners. They buy *brownfield* properties, already well developed but not delivering returns that meet the owner's expectations. They slash costs and harvest the remaining reserves, improving the returns beyond the seller's capability. Examples include Apache, Devon, and other companies who acquire the assets of large companies whose strategies focus elsewhere, such as on immature provinces.

- Strategy 4. *Accessors* penetrate government politics and bureaucracies to gain access to sovereign mineral rights. The companies deal or negotiate to gain a preferred position. The classic example is Occidental's entry, in 1966, into the Middle East, long the exclusive province of the so-called Seven Sisters (Esso, Royal Dutch/Shell, Chevron, Mobil, Gulf, British Petroleum, and Texaco). The flamboyant but successful negotiations by Occidental's Armand Hammer with Libya's King Idris simply outclassed his horrified larger competitors.

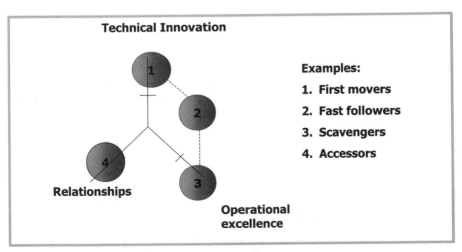

Fig. 12–4. Generic examples of E&P company strategies

Competencies are not the exclusive attributes of large companies. This three-legged framework helps explain how oil and gas companies of all sizes have survived in the face of massive migration of processes and activities to service companies. The companies probably do not have this model in mind or explicitly emblazoned on the cover of their annual reports. But they have mobilized their human resources to hire, train, or otherwise acquire people with the skills that support the three elements of the strategy.

Assessing Competencies

Not that it is any easy job to determine the competitiveness of all the competencies. Consultants offer benchmarking services to help quantify cost-competitiveness, but it often takes a large company to afford the service. For those who can, the analysis provides rankings in multiple categories or processes. For each, the detail painfully ranks each competitor (usually anonymously). Each competitor who joins in the benchmarking program recognizes its own standing.

The harsh reality is always that half the competitors lie below the average performance. In the example in figure 12–5, Competitor A used to be the low-cost manager of platform maintenance. In more recent years, Competitor C moved into the premier position, not necessarily because others' costs rose, but perhaps because of its own attention to cost reduction. In any event, "X," the client who bought this service, needs to shake up its maintenance organization—understand why its cost structure is perennially subpar and how to improve it.

Alas, for smaller companies and for everyone's appraisal of the technological innovation and relationship categories, the analysis is usually qualitative only. Still, the competencies have to be identified. Examples could include

- Technology:

    – understanding reservoir mechanics—in South Texas, for example

    – horizontal drilling

    – coal bed methane recovery—in Wyoming, for example

- Relationships:
  - contacts with foreign governments
  - crafting license applications
  - analyzing competitors' position
- Operational excellence:
  - commitment to cost reduction
  - work-process controls
  - short discovery-to-first-oil interval
  - automation/de-staffing

Identifying real, differentiated competencies is not easy work, but the effort to articulate them usually provides evidence that they at least exist. It also provides a qualitative feel for their competitive ranking.

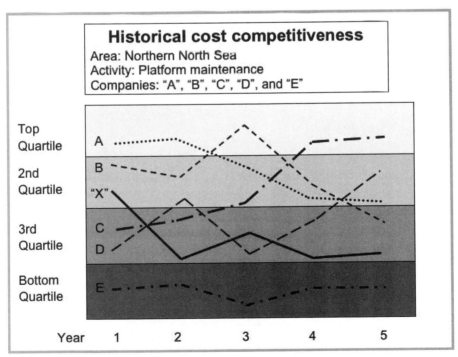

Fig. 12–5. Example of a cost-competitiveness benchmark chart

Strategy Implementation

The three-legged framework begins to explain the shift from E&P companies to service companies. E&P companies are retaining those skills essential to their strategies and shedding those that are not. At the same time, service companies are assembling or redesigning their skill sets, with or without explicitly thinking about the same three-attribute model.

Using this framework to identify or redirect strategy, management can focus resources on the competencies that matter. In addition, it provides a guide to which processes or skills can be outsourced without jeopardizing a competency important to the strategy. Finally, the framework provides a vehicle to clearly communicate in the organization the strategy and the value of the people who provide the skills that underpin it.

INDEX

J